HMS
VICTORY

Previously published in a larger format as *HMS Victory Manual Owners' Workshop Manual* in 2012 and reprinted in 2014, 2015 and 2018 This edition printed in 2020

Peter Goodwin has asserted his moral right to be identified as the author of this work.

A catalogue record for this book is available from the British Library.

ISBN 978 1 78521 688 6

Library of Congress control no. 2019945939

Published by Haynes Publishing,
Sparkford, Yeovil,
Somerset BA22 7JJ, UK
Tel: 01963 440635
Int. tel: +44 1963 440635
Website: www.haynes.com

Haynes North America Inc.
859 Lawrence Drive, Newbury Park,
California 91320, USA.

Printed in Malaysia.

Front cover: "The First Journey of Victory, 1778" by William Lionel Wyllie, 1922.

Acknowledgements

This work could not have been produced without assistance from the Commanding Officers, HMS *Victory*, The National Museum of the Royal Navy, The National Maritime Museum, The National Archives, Jonathan Coad, Chairman of the Victory Advisory Technical Committee, Dr Ann Coats of the University of Portsmouth and Naval Dockyards Society, and Nick Hall, Keeper of the Royal Armouries Fort Nelson.

I also very much thank Jonathan Falconer of Haynes Publishing for guiding me through the production of this type of book, which has given me a unique insight, allowing me to combine both my maritime historical knowledge and past marine engineering skills in this one publication.

Special thanks is extended to my long-standing friend Captain Frank Scott, a Fellow of the Nautical Institute, who, as a highly experienced master in square-rigged sailing ships, kindly oversaw my efforts to illuminate the practical elements involved in sailing the *Victory*. I also thank my colleague Martin Bibbings of Master Gunner Ltd, who has for many years generously shared his knowledge on naval gunnery and our joint experiences of re-enacting live naval gunnery firings.

Thanks are also due to Colin Mudie, John McKay, Bill Bishop, Mark Myers, Philip Chatfield, Colin Burring, MOD/Crown Copyright, iStock, Topham Picturepoint/TopFoto, and Twentieth Century Fox Film Corporation and Universal Studios and Miramax Film Corp, and iStock for their kind permission to reproduce copyright illustrations.

Finally, and most importantly, I thank my wife Katy, who has not only assisted as my Gunner's Mate with Master Gunner Ltd and sailed with me in various square-rigged ships, but has also in her own capacity as Curator of History at Portsmouth Museums and Records Service provided invaluable and supportive assistance in the creation of this book.

Peter Goodwin MPhil, I.Eng, MIMarEST
July 2011

HMS
VICTORY

Peter Goodwin MPhil, I.Eng, MIMarEST

HAYNES ICONS

Contents

INTRODUCTION	6

THE *VICTORY* STORY	12
Designing the *Victory*	14
Timber and its sources	17
Construction	18
Launching the *Victory*	25

HMS *VICTORY* AT WAR	26
HMS *Victory*'s operational career in the Royal Navy	28
The Battle of Trafalgar, 1805	34
The *Victory* after 1812	35

ANATOMY OF THE *VICTORY*	38
The decks and internal arrangements	40
Masts, yards and booms	60
Rigging and blocks	62
Sails and storage	65
Steering arrangements	66
Anchors and ground tackle	69
Ship's boats	76

VICTORY'S GUNS	78
Firepower	80
Technical data of the *Victory*'s guns	82
The velocity of the *Victory*'s guns	85
Side arms and gunnery equipment	87
Firing the guns	89
The gun drill	90
Ropes, tackles and cordage associated with guns	94
Gunpowder and its properties	96
Gunners' stores	97

GENERAL MAINTENANCE AND REFITTING	102
Hull maintenance at sea	104
Dockyard maintenance and refitting	104
Painting the ship	106
Maintenance of masts and yards at sea	109
Maintenance of sails and rigging at sea	112
Maintenance of ship's boats	114
Maintaining the guns and gun carriages	115
Repairs and action damage control	118
Capstan maintenance and repairs	118
Getting out and refitting the capstan at sea	122
Pump maintenance and repairs	123
Using the portable forge	128

SAILING THE *VICTORY*	130
Ship handling	132
Setting or taking in sails	132
Tacking and wearing ship	136
Heaving to	140
Reefing sails	141
Getting the ship under sail from being at anchor	142
Bringing the ship to anchor	146
Embarking Stores and Equipment	148
Hoisting in and getting out the ship's boats	151
Sailing performance	151

MANNING THE *VICTORY*	154
Victory's complement	156
Organisation	156
Daily routine and watches at sea	158
Commanding a first rate ship of the line	159
The seaman's view	159
Life on the lower deck of a Georgian naval warship	160

CONSERVING AND RESTORING HMS *VICTORY*	162
The problems	164
Revisiting original construction techniques	167
Dockyard facilities and tools for the job	169
Fabricating replacement parts	170
The shipwright's view – then and now	170

APPENDICES	172
List of sources and further reading	172
Appendix 1 – Glossary of Terms	173
Appendix 2 – Essential dimensions, weights, etc. for HMS *Victory*	176
Appendix 3 – HMS *Victory* visitor information	176

INDEX	177

OPPOSITE 'The starboard cherub with a blue ribbon lost a leg; the larboard cherub with a red ribbon lost an arm.' HMS *Victory*'s figurehead. *(iStock)*

Introduction

When she was launched in 1765 the *Victory* was the ultimate warship design of the Georgian age. In her day she was among the most powerful weapons systems afloat, the 21st century equivalent of an aircraft carrier or a nuclear-powered intercontinental ballistic missile.

OPPOSITE **HMS *Victory* is best known for her role as Nelson's flagship at the Battle of Trafalgar on 21 October 1805, when he defeated the combined French and Spanish fleets.** *Victory* **is preserved for all to see in No. 2 Dry Dock at the Historic Dockyard at Portsmouth.** *(MOD/Crown Copyright)*

Her Majesty's Ship *Victory*, preserved in dry dock at Portsmouth Historic Dockyard in Britain, remains a commissioned ship in the Royal Navy, serving as the flagship of Second Sea Lord and Commander-in-Chief Naval Home Command. More than this, the *Victory* is perhaps the world's most iconic preserved ship, epitomising the period of Britain's naval supremacy in the age of the fighting sail of wooden-walled fighting ships with their web of masts and rigging. Armed with smooth-bore, muzzle-loading guns firing solid iron round shot, these ships were manned by the hardy sailors we affectionately term 'Jack Tar'. As the sixth ship to bear the name *Victory* in the Royal Navy, the *Victory* is now fully restored to the form she appeared in as Admiral Horatio Lord Nelson's flagship when she fought in the victorious battle against a combined Franco-Spanish fleet off Cape Trafalgar on Monday 21 October 1805. Now colloquially called '*Nelson's Victory*', this ship is in every part upheld within the psyche of the British public and the heart of her Royal Navy. In her day the ship would have been formally titled His Britannic Majesty's Ship *Victory*, long since abbreviated to HMS, the word His or Her used according to the reigning monarch. From an international perspective she is equally important as she is representative of the classic warships of the naval powers of her era, be they French, Spanish, Danish, Dutch Neapolitan, Russian and Swedish alike.

Launched on Tuesday 7 May 1765 as a first rate ship of 100 guns, the *Victory* was the ultimate design of warship both in size, firepower and technology. In her day she was the dominant capital warship equivalent to the aircraft carrier or Trident nuclear-powered intercontinental ballistic submarine. Ships like the *Victory* remained the most potent weapons system afloat until overshadowed by the advent of steam propulsion and shell-firing guns just 40 years after Trafalgar.

The design and development of 3-decked 100-gun ships

Carrying her gun batteries ranged along three continuous gun decks, the design concept of the *Victory* originated in the first decade of the 17th century. The first '3-decker', the 56-gun *Prince Royal*, was built for King James II by the renowned shipbuilder Phineas Pett and launched at Woolwich on 25 September 1610. Modified in 1641 to carry 90 guns, the *Prince* was also the first British ship to be embellished with gilt carved works.

Next in terms of development was the *Sovereign of the Seas*, ordered by King Charles I in 1634. Also designed and built by Pett, this three-decked ship was much larger than her predecessor and launched at Woolwich on 14 October 1637. The *Royal Sovereign* (as she was later renamed) was rightfully the first ship of the line to carry the main body of its armament of 100 guns mounted upon 3 gun decks like Nelson's *Victory*. Extensively decorated and gilded, this first rate ship unfortunately proved highly

expensive and at £65,586 cost virtually the same as Nelson's *Victory* some 130 years later. Moreover, it was the unwarranted 'ship money' demanded from Parliament for this vessel that indirectly cost King Charles his head.

With the development of the tactical concept of fighting ships ranged in a line of battle firing broadsides into an equally arranged enemy formation of ships by the 'generals at sea' in the parliamentary Commonwealth, Navy ship design focused on creating steady platforms for heavy artillery. To meet this demand, warships consequently became greater in breadth and depth. They were expensive to build and no other first rate ships of 100 guns were built until the launch of the *Prince* at Chatham in 1670. Renamed the *Royal William* in 1692 and rebuilt at Portsmouth in 1719 to carry 84 guns, this ship remained in service until broken up in 1813.

The next 100-gun ship was the *Royal James*, built by Anthony Deane and launched at Portsmouth in 1671. Due to timber shortages, she was perhaps the first warship to have iron knees supporting the beam ends. Deane's ship was unfortunately burnt and

ABOVE Tens of thousands of tourists visit the *Victory* every year. More people than ever are recognising the significance of Portsmouth Naval Dockyard and HMS *Victory* in Britain's national story.
(Jonathan Falconer)

sunk by the Dutch at the Battle of Solebay on 28 May 1672. Her 100-gun successor, also named *Royal James*, was launched at Portsmouth in 1675. As a consequence of the forced abdication of King James II, the *Royal James* was renamed *Victory* in 1691 and rebuilt at Chatham three years later. In commemoration of the accession of Hanoverian King George I, this ship was renamed the *Royal George* in October 1714 and renamed *Victory* again a year later. Unfortunately, the ship was accidentally burnt in 1721 and then taken to pieces for salvage. The frames and knees and beams were set aside and reused in the construction of the fifth *Victory*, a 100-gun first rate, and this commenced at Portsmouth in 1726 with the ship finally being launched on 23 February 1737. Serving as Admiral Sir John Balchin's flagship, this *Victory*, while returning home from Portugal, foundered at night during a storm in the Channel on 5 October 1744. Over 1,000 souls, including Balchin, perished in this disaster. Lost without trace, 'Balchin's *Victory*' (as she is now colloquially named)

is thought to have come to grief off the Channel Islands on the notorious rocks called the Casquets. This location has since been proven to be false as the true wreck site of this magnificent ship was finally discovered in April 2008 by Odyssey Marine Exploration some 62 miles (100km) from her assumed last-known position.

Virtually a sister ship to Nelson's *Victory*, the 100-gun ship *Royal George* (formerly named the *Royal Anne*) launched at Woolwich on 18 February 1756. After a distinguished career and serving as the flagship of Admiral Hawke at the Battle of Quiberon Bay against the French in November 1759, the *Royal George* accidentally capsized and foundered while moored at Spithead on 29 August 1782. Admiral Richard Kempenfelt was lost in this tragedy, together with over 1,000 people, including women and children aboard the ship visiting the crew. Other 100- to 120-gun first rates and second rates of 90 to 98 guns, all 3-deckers, continued to be built after the building of the sixth *Victory*.

LEFT The *Victory* is also a shrine to Britain's most famous naval hero, Admiral Lord Horatio Nelson. New research was undertaken by the author into where in the ship Nelson actually died on 21 October 1805. The correct position is now marked by a memorial stone sculpted by Philip Chatfield. It bears the logo of the Nelson Society who, with a descendent of Admiral Samuel Hood, sponsored the project for the benefit of the ship, the Royal Navy and the Society for Nautical Research. *(Philip Chatfield)*

Chapter One

The
Victory Story

The *Victory* was the
brainchild of Sir Thomas
Slade, one of the most
innovative ship designers
of the 18th century.
His design vision, which
owed much to the styling
of contemporary French
ships, adopted a more
scientific approach,
particularly when it came
to hull shape below
the waterline.

OPPOSITE Design of the *Victory* was overseen by Sir Thomas
Slade, Surveyor of the Royal Navy from 1755–71.
(MOD/Crown Copyright)

ABOVE Nelson's
Victory is the sixth
ship to bear the name
in the Royal Navy.
(MOD/Crown Copyright)

In December 1758 William Pitt the Elder
prevailed upon Parliament to pass a bill to
build 12 ships of the line, one to be a first rate
of 100 guns. At the time, Britain was again
at war against France (the Seven Years War,
1756–63). Besides fighting in home waters,
the Mediterranean and Europe, as in previous
conflicts, this war was being fought on an
intercontinental scale encompassing America,
Canada, the East and the West Indies, India
and the Philippines and thus was effectively the
true first 'world war'.

This commitment was a response to the
need for supplementing the British fleet for the
ensuing war, British ships now being required
globally. On 13 December 1758 the Admiralty
passed a copy of the bill to the Commissioner
of HM Dockyard Chatham, authorising him to
'prepare to set up and build a new ship of 100
guns as soon as a dock is available for the
purpose'.

Designing the *Victory*

The overall design work of the *Victory* and
the 11 ships of the building programme was
undertaken by Sir Thomas Slade, who held the

office of Surveyor of the Navy from 1755 until
1771. Slade was perhaps the most innovative
surveyor of the 18th century. Prior to taking
up this prestigious appointment with the Navy
Board, Slade had been Assistant and then
Master Shipwright at the Royal Dockyards at
Woolwich and Deptford. His post as Surveyor
of the Navy had been authorised by the First
Lord of the Admiralty Admiral George Anson.
Between the years 1749 and 1769 Slade had
designed a total of 181 naval ships of war of
varying size and purpose. Having completed his
plans for the yet unnamed 100-gun ship, Slade
submitted them for approval, as the Minutes
of the Navy Board for 6 June 1759 record (see
opposite).

Slade's design for the *Victory* was very
much based on the lines of captured French
ships. Since the 1730s the French had applied
a more scientific approach to ship design,
especially with the hydrodynamic forms of the
hull shape below the waterline. This particular
factor was to enhance sailing qualities, speed
and manoeuvrability. Furthermore, by copying
French practice, Slade introduced a more
vertical stern post which improved the manner
in which a ship would answer her helm. Before

Minutes of the Navy Board, 6 June 1759

Sheer draught proposed for building the First Rate ship of 100 guns at HM Yard at Chatham pursuant to an order from the Rt. Hon. Lords Commissioners of the Admiralty 13 December last and the dimensions undermentioned viz.,

Length on the Gun Deck – 186 feet
Length of the Keel; for Tonnage –
   151 feet 3⅝ inches
Breadth moulded – 5 feet 6 inches
Breadth extreme – 51 feet 10 inches
Burthen in tons 2162. 22/94

                     Thos. Slade

To carry on the lower Deck
   30 guns of 42 pounds
To carry on the Middle Deck
   30 guns of 24 pounds
To carry on the Upper Deck
   30 guns of 12 pounds
To carry on the After Deck
   10 guns of 6 pounds
To carry on the Forecastle
   2 guns of 6 pounds

starting, the other design factors Slade had to account for were:

To make a Ship carry a good Sail.
To make a Ship Steer well, and Quickly Answer the Helm

To make a Ship carry her Guns well out of the Water.
To make a Ship go smoothly through the Water without pitching hard.
To make a Ship keep a good Wind.
A large storage capacity allowing a Ship to

BELOW Inspiration for the design of the *Victory* was based in part on the lines of captured French ships. The ship's 25ft clinker-built yawl can be seen on the dockside near the ship's bow. *(iStock)*

*operate independently from base port for
long periods.
To withstand the onslaught of enemy shot in
order to protect the Ship's own gun crews.*

On 7 July 1759 the Navy Board sent an official
instruction to the officers at HM Dockyard
Chatham; the letter read as follows:

*By the Principal Officers
and Commrs. of his Majys. Navy.*

*Pursuant to the order from the Rt. Hon.
the Lords Commrs. of the Admiralty dated
the13th Decr. 1758 and 14th of last month,
these are to direct and require you to cause
and be set up and built at your yard a new
ship of 100 guns agreeable to the Draught
herewith sent to you and of the Dimensions
set down on the other side hereof, and
you are to forthwith to prepare and send
us in due form an Estimate of the Charge
of Building and fitting for sea the said Ship,
and providing her with Masts, Yards, Sails,
Rigging and Store to an eight months'
proportion. For this shall be your warrant.
Dated at the Navy Office the 7th July 1759.*

*(signed)
Richd. Hall Tho. Slade G. Adams
Th. Brett.*

The Commissioner at Chatham was Captain
Thomas Cooper, who was in poor health at the
time and had to retire the following year. The
Master Shipwright overseeing the construction
of the *Victory* was John Lock.

John Lock had previously served as the
Assistant Master Shipwright at Portsmouth
between 1742 and 1752. Lock's brother
Pierson, it is noted, was the Master Shipwright
at Portsmouth during that period and died in
1755. Their father, also named John, had been
Master Shipwright at Plymouth in 1705.

The design of any 18th-century warship
necessitated that it had to be fit for purpose. In
the case of the *Victory* its desired function had
to be a 'stable manoeuvrable floating fighting
gun platform'.

The keel of the *Victory* was laid down in the
Old Single Dock at Chatham on 23 July 1759.

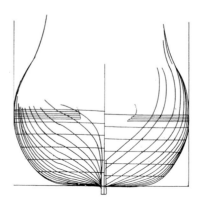

The site of this historic dock still exists today,
albeit the dock has been extensively modified.
The dock actually lies on an east–west axis
with its dock gate entering the River Medway to
the west. Of note, the orientation of this dock
in relation to the elements and the sun would
indirectly affect the construction of the ship.

As always in times of war, the Royal
Dockyards like Chatham were working at
maximum capacity with many craftsmen
employed. To increase the workforce to meet
the demands of the war the Navy Board sent
the following letter to HM Dockyards in February
1759: 'These are to direct you to cause all
possible dispatch to be used in cleaning,
refitting and storing of the ships of the line at
your Port, as well as the frigate, it being of the
greatest consequence that not a moment's time
should be lost in getting them ready for service.'

Chatham was no exception, and in all
some 150 men were initially employed to
construct the *Victory*. With the elm dock
blocks laid along the axis of the dock, the
keel of this yet unnamed ship was laid down
on Monday 23 July 1759. This occasion was
attended by Prime Minister Pitt and Admiralty
representatives who had travelled down from
London. After witnessing the keel being laid,
all attended a banquet. The year 1759 was
significant in that Britain attained a series of

ABOVE AND RIGHT
***Victory* to the original
draught as built.**
(Peter Goodwin)

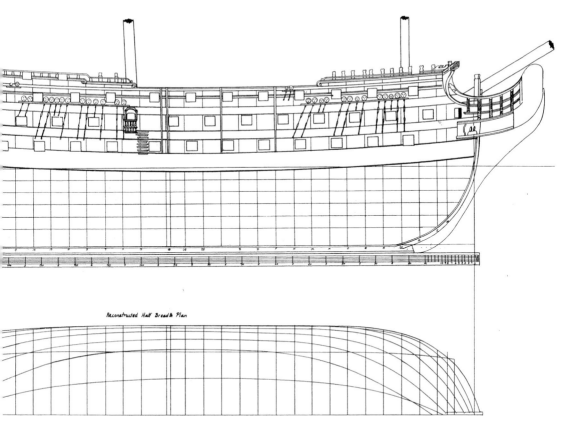

Reconstructed Half Breadth Plan

victories that formulated the turning point of the Seven Years War, the French being defeated at Quebec, Minden, Lagos and Quiberon Bay. Consequently, this year would later be dubbed the *Annus Mirabilis*.

Timber and its sources

Some 300,000cu ft of timber was used in the construction of the hull of a first rate ship like the *Victory* before conversion into separate components. This amounts to 6,000 trees taken from 100 acres of woodland, 90% of which comprised British oak. The best oak came from the Weald forests of Kent and Sussex, which could easily be shipped on barges down the River Medway to the dockyard at Chatham. Oak from Kent and Sussex was particularly good as the native heavy weald clay soil promoted slow growth producing a dense, robust timber.

RIGHT Most of the timber used to build ships like the *Victory* was oak, from the Weald forests of Kent and Sussex. *(Katy Ball)*

As well as British elm, other timbers comprised pine, fir and spruce, most of which was imported from the Baltic States and Riga, together with straight oak from Danzig (modern Gdańsk). Beech was also used, although this species of timber had some failings. Pine for the *Victory's* great lower masts during this particular period would still have been imported from New England, where single trees, some 130ft long and 3ft in diameter, to form 'pole' masts could be found. After the American Declaration of Independence in 1776, this source of mast timber was restricted for the duration of the resultant war. Other materials needed in great quantities for the ship's hull were iron nails, spikes and bolts and also copper bolts of up to 2in in diameter and 10ft in length. In addition to these items, vast quantities of iron or copper roves (washers) were required for use with the bolts.

LEFT The Rother (rudder) stern post and inner post. *(Peter Goodwin)*

Construction

The keel forms the backbone of the ship. It is made up in seven sections from British elm, which was used for two reasons: its inherent durability when immersed in seawater for very long periods; and its irregular grain structure, which can withstand having many bolts driven into it without splitting. This second factor is highly important as there were many bolts driven through the keel to join the floors of the ship's main timbers (or ribs). Recent evidence suggests that the elm for the *Victory's* keel was sourced near Portsmouth and shipped to Chatham. If this was the case, it certainly provides an insight into the logistics of inter-dockyard support regarding material procurement and transfer.

The keel

Laid down on Monday 23 July in 1759, the keel measures 151ft 3.625in in length, 21in in depth throughout its length, 21in in breadth amidships, tapering to 16in afore and 18in abaft. The seven sections forming the entirety of the keel are joined together with vertical scarphs about 5ft in length, bolted together with eight copper bolts ¼in in diameter driven horizontally. A rabbet (rebate) 5in deep was cut horizontally along the top edge of the keel to receive the inner

edge of the lowermost hull plank, called the garboard and made of well-seasoned straight oak. Next, fastened upon the upper face of the keel is the rising wood (or hog) made of some eight lengths of oak 2ft in width and some 6in in depth. The function of the rising wood is to form a land for the transversely placed floors of the frames of the ship. The upper face of the hog is fashioned with a form of 'cross-halving joints' to receive the floors.

BELOW Ironwork securing heels of the stern counter timbers and wing transom end to the ship's side. These may have been fitted as early as c.1810, but more probably during the rebuild of 1814–16. *(Jonathan Falconer)*

The stern post

Rising at the after end of the keel and forming the aftermost boundary of the ship's hull is the stern post made from the bowl of a large single oak tree of at least 100 years old. This substantial timber measures 31ft 3in in height, 2ft 2in square at the head, 3ft in breadth fore and aft at the heel and 18in wide athwartships at the keel. The heel of the stern post is tenoned into the top edge of the keel. The back edge of the main post is bearded to give the rudder 45 degrees of helm. The stern post is supported on its fore side by an inner post made of oak. Besides providing a means from which vertically to hang the ship's rudder, the stern post also serves to support the great beam called the wing transom upon which the entire stern structure and counter timbers are supported above the level of the gun deck.

The inner post

Fitted to the fore side of the stern post and terminating under and supporting the wing transom, this timber measures 1ft 4in fore and aft at its head and 1ft 7in at its heel. Like the main post, it is tenoned into a mortise cut into the top edge of the keel. Both the main and inner posts are additionally secured and supported at their heels by copper fish plates bolted through with copper bolts. Both posts are jointly supported at their fore side by a series of horizontal timbers bolted together, collectively forming a large singular inverted knee colloquially called the dead wood. Made from oak, the dead wood also provides a land onto which the heels of the cant (angled as opposed to transverse) frames forming the shape of the after body of the ship as it curves inward to the middle line. The copper bolts driven down through the dead wood are 2in in diameter, varying from 6ft to 15ft in length.

The wing transom

Mentioned previously, this great oak beam is over 30ft in length, moulded some 1ft 2½in at its centre of length and 2ft 3in at its extremities and bolted to the head of the stern post with two bolts of 1³/₈in diameter. Fitted below this is the deck transom, which forms the aftermost beam of the lower gun deck; this is moulded 1ft 3in and fastened with two 1¼in diameter bolts.

Transom beams

Next, forming the round tuck of the stern, seven transom beams are fitted filling the space between the deck transom and the aftermost frame called the fashion piece. The transom beams are supported internally with transom knees or sleepers, although these were not installed until internal planking was fitted.

Note: during construction the stern post and inner post were actually fitted up upon the after end of the keel as a large sub-assembly together with the wing transom and transom beams.

The stem post

Rising in a curve at the fore end of the keel and forming the foremost boundary of the ship's hull is the stem post, made from three sections of compass oak taken from trees of at least 100 years old. The height of the stem post is given as 40ft 10in and its thickness (moulded) 1ft 9in. Its breadth athwartships at the head is 2ft 6in and breadth at the heel 18in. The scarphs joining the three sections of the stem are 4ft 6in in length. Each scarph is bolted together with eight copper bolts 1¼in in diameter. The heel of the stem post is joined into the fore end of the keel with a complex scarph called the boxing. This intricately cut scarph is bolted together with eight copper bolts 1¼in in diameter and driven horizontally. Fitted to complete the fore foot of the keel and its union with the stem post is the gripe, made from a single baulk of oak bolted radially into the stem with copper bolts 1⅛in in diameter. The gripe is further secured and

BELOW Sternson, transom beams and sleepers. The heavy angled timbers are sleepers that give support to the transom beams.
(Jonathan Falconer)

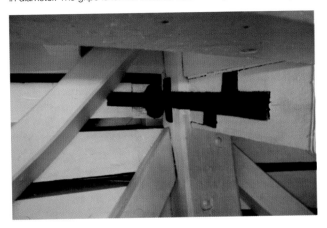

RIGHT Bowsprit and
the supporting knight
heads, which are
formed by the
innermost hawse
pieces.
(Jonathan Falconer)

sided 1ft 2in at their heels. Their uppermost
end equally forms a yoke to retain the ship's
bowsprit in position.

The hawse pieces

Made from oak, these comprise four vertical
timbers fitted either side forming the main
shape of the bluff bow of the ship and are
designed to force the ship's head through the
sea. The foremost is sided 1ft 9in and 1ft 8½in
at its head, the second and third are sided 1ft
8in and 1ft 7½in at their heads and the fourth
is sided 1ft 9in and 1ft 8½in at its head. All are
sided 6in at the heel. An 18in-diameter hole is
cut through the middle of each pair to form a
hawse hole through which the anchor cables
enter the ship. The hawse pieces are fastened
together with 1¼in-diameter bolts with one
above the hawse holes and three below.

The ship's timbers or frames

The main or square frames were made from
selected pieces of compass oak and form the
cross-sectional body of the ship. Complex and
important in shape, the manufacturing process
to generate the frames started in the dockyard
mould loft, a large, open room with clear floor
space. Here the shipwrights transferred the
shapes from the ship's body plan and marked
(scrieved) them out in full scale on the mould
loft floor. From this templates made from deal
battens were used to mark out the shapes
for cutting the timber components to form
the overall frame from keel to the topside of
the ship. Each frame comprises two halves
fayed and bolted together. As the faces of
each frame form the side of the gun ports and
all guns need a working room of 7ft 6in, the
spacing of the ships timbers along the keel is
critical. This spacing, called the 'timber space
and room', for the Victory is given as 2ft 9⅜in.
This distance accounts for one main frame
and one filling frame, the single frame being
considered as a single half frame. Whether
main or filling, each frame is made up with a
series of individual pieces comprising floors,
1st futtocks, 2nd futtocks, 3rd futtocks and
4th futtocks forming the underwater body of
the ship and extended with top timbers and
lengthening pieces to form the ship's sides. All
floor timbers (a total of 53 in number) are sided

ABOVE Starboard
anchor cable, hawse
holes and cheek.
Note the middle gun
deck chase port with
half-port lids and
washboard.
(Peter Goodwin)

supported by use of thick copper fish plates
and horseshoe plates bolted through with
horizontally driven copper bolts.

The apron

Acting like an inner stem post, the apron is
made from compass oak 1ft 2in thick and 2ft
6in broad, the scarphs giving shift to those of
the stem post and 1ft 10in long.

The knight head timbers

These are fitted vertically either side of the
apron and provide additional transverse bracing
to the stem-post structure. They are made of
oak 1ft 5in and 1ft 4½in square at the head and

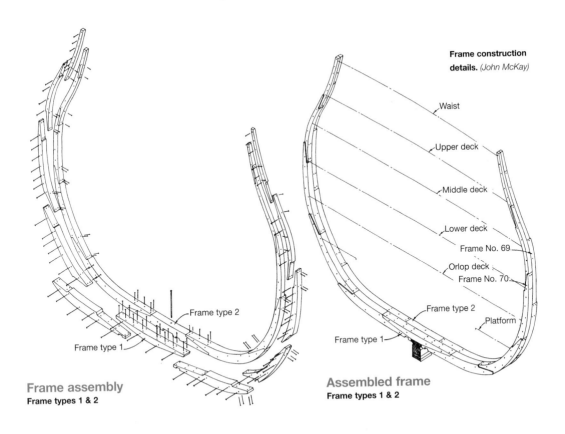

Frame construction details. *(John McKay)*

Waist

Upper deck

Middle deck

Lower deck

Frame No. 69

Orlop deck

Frame No. 70

Frame type 2

Platform

Frame type 1

Frame type 2

Frame type 1

Frame assembly
Frame types 1 & 2

Assembled frame
Frame types 1 & 2

LEFT Oak futtock riders, ceiling and strakes of 'thick stuff'. 'Ceiling' is the term for the internal planking in the lower part of the hull (originally spelt with an 's'), which seals the inside of the frames.
(Jonathan Falconer)

was transported to the building dock. The scarphs of the various components are bolted with three square bolts of 1½in in diameter complete with chocks. Each frame is retained in its shape by temporary battens called cross spiles and erected upon the keel and fastened to the floors. Each component is raised up with portable capstans called crabs, which are set on the dock floor, using a series of ropes through pulley blocks traversing along a long heavy line supported by sheer legs at each end of the dock. This line is set up at a desired working height over the length of the keel. The process of raising up the frames commenced afore and abaft the mid-length point. The crabs were set on the dock floor as required.

ABOVE Head of sternson, standard and wing transom in the gun room amidships.
(All pictures this spread Jonathan Falconer)

The keelson

Virtually serving as an inner keel running fore and aft directly over the keel and hog, this comprises a series of oak timbers 1ft 7in sq in section with horizontal scarphs. The function of the keelson is to lock down all the frame floor timbers and cross chocks in position. The after end of the keelson is formed into a sternson knee, which, like an inverse bracket, supports the stern post and inner post. The fore end of the keelson, called the stemson, rises steeply over the rising wood to support the apron and stem post.

RIGHT Upper end of standards (inverted knees) supporting orlop beams below. This area, part of the breadroom, was originally not decked over, the planking being added in the 20th century.

The cant frames

Fitted to form the hull and curving in towards the middle line in the fore and after body of the ship, these frames, like the main or square frames, comprise floors, futtocks, top timbers and lengthening pieces. All cant frames are set at a series of diminishing angles to the keel to avoid reducing their siding (width) by trimming their outer faces to create the hull curvature required. Having no true land, their heels are bolted direct onto the dead wood and rising wood afore and abaft.

Note: the siding and moulding specifications for individual main, filling and cant frame components, i.e., futtocks, top timbers, lengthening pieces and the short timbers over gun ports, etc., range between 1ft 4in and 1ft 2½in depending on their disposition.

1ft 3in, fitted individually and bolted to the keel and rising wood with three 1½in-diameter bolts driven through the keelson. In compliance with the ship's body plan, all frames are individually labelled, using letters in the fore body of the ship and numbers abaft. The midpoint of the framing system is called the dead flat and given the symbol ¤. The constituent parts of the rest of each frame, cross chocks, futtocks and top timbers, were manufactured and fitted together forming a single component, each of which

The 100-gun ship under construction was eventually named *Victory* in October 1760. Although the sixth ship in the Royal Navy to bear this name, the decision did not come lightly. Her predecessor, the fifth *Victory* (often called *Balchin's Victory*), equally a first rate of 100 guns, had foundered with all hands in 1744. However, despite this tragedy, the choice to give the name *Victory* was very much influenced by the events of 1759, the 'Year of Victories', the *Annus Mirabilis*. The original elaborate figurehead, which was to adorn the ship, portrayed these events. Suffering rot, this figure and its supporters had to be removed during the *Victory*'s 1800–3 refit and was replaced with a less costly modest figure.

Note: a model of the original figurehead is contained in the collections of the National Maritime Museum. A quarter-scale model is likewise displayed in the *Victory* Gallery of the National Museum of the Royal Navy, Portsmouth.

Under the pressure of war, it had been planned to complete the ship within 30 months, quite an undertaking at that time. However, as the hostilities ended with the signing of the Peace of Paris on Thursday 10 February 1763, the urgency to complete the *Victory* was less critical. Consequently, the workforce at Chatham was considerably reduced, with sections being laid off and leaving just 1,325 in full employment.

Once most of the framing was complete the ship was left to stand 'in frame' to allow the structure to settle. This provided the time for trimming adjustment and seasoning, the whole being retained by longitudinal battens called ribbands. It has often been remarked that the *Victory* attained her much-seasoned strength to her long standing in frame. However, while this largely mythical concept may have been a contributory factor, in reality her time 'in frame' was governed more by events rather than practice, mainly because the Seven Years War had ended and there was no urgency to complete the ship. The other factor reiterated above lay with finances. Of particular note, the practice of having all ships left in frame for at least a year did not become official policy

ABOVE Oak floor and futtock riders and ceiling and strakes of thick stuff wrought over the wrung heads and floor heads. The hold with its internal planking extends from the keelson and floor timbers up to the beam clamp that supports both the orlop and lower gun deck beams.

CENTRE Elm limber board and keelson. Horizontal hand holes to aid removal can be seen.

LEFT Main and filling frames exposed between strakes of thick stuff over the third futtock heads.

until 1771, by which time it had been found that ships planked over green frames were rotten and condemned after four to five years. Although this problem is believed to have been the result of either bad building practice or the urgent need to complete ships, ongoing research equally reveals that this deterioration may have been influenced to some degree by climatic change, winters not being cold and dry enough and warm, damp summers (a problem that has revealed itself in the *Victory* today).

In summary, the initial construction work of the *Victory* was overseen by the Master Shipwright John Lock who, in failing health, unfortunately died in 1762. Lock was succeeded by Edward Allin who had been serving as the Master Shipwright at Sheerness Dockyard since 1755.

The internal planking

Using a reduced workforce, the *Victory*'s hull was slowly completed over the next years in plank, beams and knees together with a multitude of necessary internal structural stiffeners and pillars. First, heavy internal stringers of straight oak were wrought fore and aft to provide longitudinal strength and bind the frames together. This comprised the footwaling adjacent to the keelson and various bands of thick stuff covering the frame joints of the futtocks and timbers, the thickness of which varied between 5in and 7in. Next, longitudinal deck clamps up to 9in thick to support the ends of deck beams were installed and bolted to inner faces of the frames. Of note, deck clamps are not to be confused with what were to become known colloquially as beam shelves under construction systems introduced by Naval Surveyor Sir Robert Seppings much later. Following the clamps was the spirketting, which was fitted along each deck level and worked over the beam ends up to the gun-port sills. All of the above provides longitudinal strength to the hull. The framing in the hold was lined closed with its ceiling (originally termed as the sealing). Likewise, the internal faces of the frames in the spaces formed between the upper edges of the spirketting and lower edge of the deck clamps were planked with 'quick work', usually short lengths of oak or pine worked between the gun ports. The size of all these bands of internal planking varies according to their disposition and function.

The deck beams

Beams were next fitted and secured down upon their clamps and fastened to the ship's side with two timber brackets at each end; those in the vertical plane called hanging knees, those fitted horizontally, lodging knees, the number and sizes of diameter of these bolts varying according to their location. During build the *Victory* was fitted with curved wooden hanging knees cut from trees selected with natural curved grain direction to maximise strength where applied. In time, 40% of these would be replaced with wooden chocks braced with iron. Before the actual deck planking was laid, the spaces between the beams were furnished with a lattice of intercostal timbers; carlings running fore and aft, and transverse ledges between.

The external planking

After planking up the ship's bottom with straight oak some 4in thick and with strakes of thick elm and beach boards closer to the keel, elm work commenced on planking the ship's bulwarks above the waterline. First, three heavily built bands of planking called wales (lower, middle and channel) were wrought longitudinally between the gun ports, the lowest (or main wale) near the waterline being some 10in thick. Besides binding the

BELOW Breast hook and vertical apron at the fore end of the middle gun deck.
(Jonathan Falconer)

frames together, in principle the wales serve the same function as the hogging trusses used in ancient Egyptian shipbuilding to reduce the unavoidable and inherent hogging and sagging effect; the upward or downward bend at the ends of a ship's hull when afloat or riding in a sea. Despite the constructive counter-measures taken, most wooden ships 'hog' 6in as soon as they are launched and possibly further 'creep' to 12in. Between the wales the hull was fitted with short lengths of oak planking between the gun ports, the thickness of these planks varying accordingly from 6in above the main or lower wale to about 3in at the topside of the bulwarks. The stern with its counters and transoms was next planked in and completed with the stern lights (windows) and open stern galleries (later removed) and the quarter galleries; all were adorned with carved works. Other works comprised building the transverse beakhead bulkhead, closing off the fore end of the upper gun deck complete with roundhouses. Finally, the figurehead was fitted in position together with the carved trailboards and supporting sweeping head rails.

Launching the *Victory*

The *Victory*'s hull was completed on St George's Day, Tuesday 23 April 1765, and subsequently Edward Allin, the Master Shipwright, wrote to the Navy Board on Sunday 28 April stating, 'His Majesty's Ship *Victory* building in the Old single Dock will be ready to launch the ensuing Spring tydes'. Invitations for the launch were sent to the Prime Minister William Pitt (the Elder), members of his Cabinet and members of both Houses of Parliament, who all travelled by coach to Chatham for the ceremony. The *Victory* was duly 'launched' on Tuesday 7 May 1765. The *Victory* was not actually launched in the broadest sense, but simply 'floated off' the dock. During this period, all first rate ships were actually built within an enclosed dry dock as they were too large to construct in the conventional manner on a slipway. Her overall cost including masts, yards, rigging and stores, etc. amounted to £63,176 3s 0d. Today, this figure equates to approximately £46.5 million.

ABOVE Quarter galleries, gingerbread, cove, taffrail and lanterns. Note the mizzen mast with the topmast and topgallant mast struck.
(Jonathan Falconer)

Chapter Two

HMS *Victory* at War

As Nelson's flagship at the Battle of Trafalgar in 1805, the *Victory* has become linked to the memory of Britain's greatest naval hero. However, in 1799 she was almost scrapped, reprieved only when the 98-gun *Impregnable* foundered, her loss creating an urgent need for a first rate ship of the line.

OPPOSITE 'The Naval Battle of Trafalgar on October 21st 1805', by William Clarkson Stanfield. *(Topfoto)*

The American Revolutionary War, 1775–83

In 1778 France, Spain and Holland entered the War of American Independence in support of the colonists. Consequently, in February that year the *Victory* was put into commission for sea service as flagship for Admiral Augustus Keppel commanding the Channel Fleet. At this stage, the ship's lower gun deck armament of 42-pounder guns was replaced with less cumbersome 32-pounders. On 23 July Keppel fought an indecisive battle against d'Orvilliers's French fleet off Ushant, during which the *Victory* suffered few casualties or damage. Returning home, the 42-pounders were replaced in the ship. Over the next three years the ship served as flagship for Admirals Hardy, Geary, Drake and Parker. In March 1780 *Victory* was taken into dock to have the bottom of her hull sheathed in copper plating, a new but expensive innovation introduced to combat the problem of the shipworm *teredo navalis*. Because the copper sheathing also deterred marine growth and barnacles it also improved the ship's speed.

Now flying the flag of Admiral Richard Kempenfelt, the *Victory* fell in with a French fleet from Brest off Ushant on 13 December 1781. The French were escorting a convoy of troopships bound for the West Indies. Although Kempenfelt's squadron was numerically inferior, his ships managed to capture the entire convoy under the noses of the escorting fleet. Serving as flagship of Admiral Richard Howe in October 1782, the *Victory* took part in an action off Cape Spartel and supported the relief of Gibraltar. With the close of the war the *Victory* was recalled in March 1783 and went into dock and refitted in Portsmouth at a cost of £15,372 19s 9d. During this period her iron hull bolts below the waterline were replaced with those of copper and other non-ferrous alloys to avoid the electrolysis reaction resulting from her copper sheathing. Her quarterdeck armament of 6-pounder guns was changed to short 12-pounders. Her sides, originally painted 'bright' with rosin down to the top of the lower deck gun ports, were now painted a

ABOVE Vice Admiral Horatio, Lord Nelson, used the *Victory* as his flagship at the Battle of Trafalgar in 1805, and whose name is most closely associated with the ship. From a portrait of 1801 by Sir William Beechey (1753–1839). *(City of London/TopFoto)*

HMS *Victory*'s operational career in the Royal Navy

When the *Victory* was launched in 1765 Britain was no longer at war and, therefore, after initial sea trials, the ship was immediately laid up in 'ordinary (reserve)' and spent the next thirteen years moored in the River Medway.

dull yellow ochre, including the external faces of her gun-port lids; the ship's side below this point was painted black. In 1787 the ship underwent a 'large repair' at a cost of £37,523 17s 1d. During this refit the ship was further strengthened by fitting breadth, middle and top riders, which provided additional internal bracing to her hull between the orlop deck and the upper gun deck to counteract the inherent hogging and sagging effects on an ageing hull, now 22 years old. Recommissioned in 1789, the ship again served as flagship to Admiral Howe and flagship to Admiral Hood in 1790.

The French Revolutionary War, 1793–1802

The storm clouds of war were gathering over Europe. In 1789 France was hurled into a bloody revolution with the execution of King Louis XVI. In January 1793 France declared war on Britain, Spain and Holland, plunging Europe into a devastating war that was to last a quarter of a century. The *Victory* immediately became flagship to Admiral Samuel Hood, Commander-

RIGHT Commanding a 13-strong squadron of ships of the line, Rear Admiral Richard Kempenfelt (1718–82) in HMS *Victory* defeated a numerically superior French force at the Battle of Ushant on 12 December 1781, during the American War of Independence. Later, as a junior flag officer, he drowned with over 800 other men on the 100-gun HMS *Royal George* when she sank at anchor at Spithead the following year. Engraving by H. Robinson after a portrait by Tilly Kettle (1736–86). *(TopFoto)*

ABOVE The *Victory* in 1793, going down the English Channel, outward-bound with her squadron for the Mediterranean, where she was Admiral Hood's flagship at the Siege of Toulon and the invasion of Corsica. She is shown flying the flag of Lord Hood as Vice Admiral of the Red. From a painting by Monamy Swaine. *(Topfoto)*

in-Chief of the Mediterranean Fleet blockading the French naval base of Toulon. During this time the *Victory* captured and destroyed several French ships and in July participated in the siege of San Fiorenzo, Calvi and Bastia on the island of Corsica. A promising young captain named Horatio Nelson led the daring amphibious assault upon Calvi. It was at Calvi that Nelson lost the sight in his right eye. In 1794 the *Victory* returned to Plymouth where Hood, his health shattered, lowered his flag. After a minor refit the *Victory* sailed again to the Mediterranean as flagship of Rear Admiral Robert Man. In July 1795 the *Victory* led the offensive in the unsuccessful action off Hyères during which Vice Admiral William Hotham failed fully to engage the Toulon fleet. Nelson, commanding the 64-gun *Agamemnon*, played a minor but distinguished role in this battle. The failure to break the Toulon fleet and the lack of safe ports led the British to abandon maintaining a fleet in the Mediterranean. After a brief command under Vice Admiral Robert Linzee, Admiral John Jervis hoisted his flag in the *Victory* in December 1795. By this point Spain had entered the war on the side of France. On 14 February 1797 Jervis's squadron of 14 ships of the line fell in with a Spanish fleet comprising 27 ships off Cape St Vincent and commanded by Admiral Don José Cordoba. A decisive victory ensued mainly due to the initiative of Nelson, who, now commanding the 74-gun ship *Captain*, quit the line of battle and moved to prevent the Spanish making good their escape. In doing so, he engaged, boarded and captured the great Spanish

112-gun ship *San Josef*. More than this, Nelson then used this ship as his 'patent boarding bridge' and captured the neighbouring 80-gun *San Nicholas*. As a result of this action, Sir John Jervis was given the title Earl St Vincent and Nelson earned a knighthood and was promoted to rear admiral.

In October 1797 the *Victory* returned to Britain and was surveyed at Portsmouth. Now 32 years old and somewhat battle worn, she was sent up to Chatham to await her fate. On 8 December the *Victory* was listed as unfit for service and ordered to be converted into a hospital ship for French prisoners of war. This decision virtually guaranteed her demise and final journey to the breaker's yard. Fate intervened when on 8 October 1799 the first rate 98-gun *Impregnable* foundered off Chichester harbour. This loss created an urgent need to retain a first rate ship in the Channel Fleet in a time of war. Consequently, the *Victory* was given a new lease of life and taken in hand at Chatham Dockyard, where the survey revealed that she was 'in want of a middling repair' at an estimated cost of £23,000.

Refitting commenced in late 1800, and closer examination revealed that parts of the hull needed rebuilding, over 60% of the wooden knees required refastening or replacing and many of the gun-port lids needed refitting. In compliance with new improvements authorised by the Navy Board in 1798, the open stern galleries were removed and the entire stern was 'closed in'. The ornate works on the stern were also reduced in keeping with expense restrictions authorised in 1798. The open bulwarks were raised on the quarterdeck and possibly her forecastle. Her magazines were lined with copper, conforming to current practice. Two additional gun ports were cut on the lower gun deck. The original ornate figurehead, found to be rotten, was replaced with that of a simpler design, comprising a badge of the royal coat of arms and Hanoverian arms surmounted with a crown and cherub supporters. The ship was also furnished with many new innovations, resulting from the experiences of war since 1793. These comprised extra shot lockers on each gun deck, cisterns to flood the magazines and a speaking tube providing communication between the ship's steering wheel and the tiller down in the gun room. The lower pole masts (each made from a single tree) were replaced with composite-built masts bound with iron hoops. The triangular mizzen sail set on a lateen yard was replaced with a quadrilateral mizzen sail set on a gaff and boom. The external sides of the *Victory* were now painted with alternate black and yellow bands, although the gun-port lids remained yellow at this stage. Internally, the sides of the ship were painted yellow or whitewashed, rather than the red ochre paint used previously.

On 5 February 1801 Prime Minister William Pitt the Younger was forced to resign as a result of political pressure over Catholic emancipation. By March the French Revolutionary War had financially exhausted Britain, France and Spain. In order to destroy Napoleon's Northern League against Britain a fleet under Admiral Sir Hyde Parker, with Nelson as second-in-command, sailed to break the Danish allegiance. The resultant Battle of Copenhagen ensued on 2 April. It was during this hard-fought action that Nelson 'turned a blind eye' to Parker's signal to withdraw and negotiated a truce with the Crown Prince. This battle thwarted Napoleon's ambitions and the new Government led by Henry Addington negotiated peace with France, the treaty of which was signed on 27 March 1802, and thus ended the French Revolutionary Wars.

The Napoleonic Wars, 1803–15

The peace was fragile and short-lived. The *Victory* was finally completed and undocked on 11 April 1803, her 'great repair' costing a total of £70,933, 66% more than the original estimate. To reduce weight on her old hull and improve firing rate, her heavy 42-pounder guns were replaced with lighter and more manageable 32-pounders, with

BELOW Thomas Masterman Hardy (1769–1839) served as Flag Captain to Nelson and commanded HMS *Victory* at the Battle of Trafalgar on 21 October 1805. In the hour of victory Nelson was shot and fatally wounded as he paced the quarterdeck with Hardy. Engraving by H. Robinson after a portrait by Richard Evans (1784–1871).
(Topfoto)

an extra two placed at the new gun ports. The six 18-pounder carronades previously mounted on the poop deck were removed, although the lightly boarded bulwarks remained. Recommissioned under her new commander Captain Samuel Sutton on 9 April, the *Victory* sailed for Portsmouth on 14 May. Just two days later hostilities against France reopened with immediate threat of invasion by Napoleon Bonaparte. Nelson was appointed as Commander-in-Chief Mediterranean Fleet. The *Victory* had provisionally been appointed as flagship for Admiral Cornwallis, commanding the Channel Fleet already blockading the French fleet in Brest. Despite this, Nelson, now a vice admiral, hoisted his flag in the *Victory* on 18 May. Unfortunately, the *Victory* was not yet ready for sea, so urgently needing to get out to the Mediterranean Nelson lowered his flag on 20 May and took passage to the Mediterranean in the 36-gun frigate *Amphion*, commanded by his old friend Captain Masterman Hardy. Before leaving Spithead Nelson told Captain Sutton that should Admiral Cornwallis not need the *Victory* Sutton was to proceed to the Mediterranean and join him there. Cornwallis declined the ship, although his hall chairs did not get transferred and have remained in the *Victory* to this day.

When Sutton reached the Mediterranean

Fleet off Cape Sicie, Nelson, taking Hardy with him as his Flag Captain, transferred his flag into the *Victory* on 31 July, while Sutton, as ordered, took command of the *Amphion*. From this point Nelson, Hardy and the *Victory* become historically synonymous. For the next 18 months the *Victory* remained on station blockading the French in Toulon and periodically refitting at the anchorage in Agincourt Sound, Corsica. It was while here on 18 May 1805 that two frigates, Nelson's 'eyes of the fleet', suddenly appeared signalling that the French fleet, commanded by Admiral Piere Villeneuve, had escaped out of Toulon. The *Victory* immediately weighed anchor and the 'great chase' began, leading Nelson's squadron eastward to Alexandria. Not finding the French there, Nelson sailed west and into the Atlantic and back, passing through the Strait of Gibraltar on 4 May. As part of the invasion plan Napoleon sent Villeneuve to the West Indies as a ruse to draw off the British fleet defending the English Channel and to enable Napoleon to land his invasion troops prepared at Calais and Boulogne. Nelson in the *Victory* followed in hot pursuit across the Atlantic and back, finally running Villeneuve to

ground at the Spanish port of Cadiz. Nelson left his long-term colleague Admiral Cuthbert Collingwood to blockade what was now a combined Franco-Spanish fleet. Nelson, much exhausted, sailed for Portsmouth in the *Victory*, arriving there on 18 August. After a brief rest at Merton, Nelson sailed from Portsmouth on 15 September and rejoined Collingwood off Cadiz on 28 September, celebrating his own 47th birthday on arrival.

Following a succession of plans thwarted by the overall stranglehold of blockade by various naval squadrons as well as Nelson, Napoleon had already abandoned the invasion of Britain in July of 1805. Faced with a new coalition

ABOVE Nelson lies fatally wounded on the *Victory*'s quarterdeck, shot by a French sharpshooter. **From a painting by Daniel Maclise (1806–70).** *(World History Archive/TopFoto)*

The fleets at the Battle of Trafalgar

British Fleet	Combined Franco-Spanish Fleet
27 line of battle ships	33 line of battle ships
4 frigates	7 frigates
1 schooner	3 brigs
1 armed cutter	
guns: 2,148	guns: 2,632
men: 17,000	men: 30,000

between Britain, Austria, Russia and Sweden (and shortly afterwards Prussia), Napoleon ordered his troops to march east to open a new theatre of war against the Austrians and this alliance. To this end, Napoleon ordered Villenueve to sail with the Combined Fleet into the Mediterranean and Italy to provide naval support to this campaign. On 18 October Nelson's watching frigates signalled that the enemy were weighing anchor to make good their escape and sailed to enter the Mediterranean through the Strait of Gibraltar. Unable to shake off the pursuing British fleet, Villenueve turned back to Cadiz and inevitable battle.

The Battle of Trafalgar, 1805

At daybreak off Cape Trafalgar on Monday 21 October 1805 the British fleet formed into two lines, one led by Nelson in the *Victory*, the other by Collingwood in the 100-gun *Royal Sovereign*, and sailed at right angles towards the enemy line. Battle commenced around 11:45am with Collingwood breaching

BELOW 'The Death of Nelson' by Arthur William Devis, 1807.
(Topfoto)

the rear of the Combined Fleet. Shortly after, the *Victory*, driving into the van of the fleet, fired a devastating broadside into the stern of Villenueve's 80-gun flagship *Bucentaure*. Next she went alongside and engaged the 74-gun *Rédoutable*, commanded by the very able and tenacious Captain Jean Jacques Lucas. At about 1:15pm, when the fighting between the two ships was at its fiercest, Nelson was fatally shot by a French marksman and carried below, where he died of his wound at about 4:30pm. As he lay dying, the Franco-Spanish fleet was routed and the Royal Navy won its most glorious victory: 17 ships were captured and one, the French 74-gun *Achille*, blew up at about 5:30pm.

A tremendous storm followed the battle, during which many of the captured ships were lost and much damaged. The *Victory*, with her mizzen mast and topmasts shot away, was finally towed by the 98-gun *Neptune* into Rosio Bay, Gibraltar, on 28 October. Having made good her repairs and fitted jury masts, the ship sailed for Britain on 28 November carrying the body of the fallen hero Nelson. She arrived at Portsmouth on 4 December.

ABOVE Carrying
Nelson's body, the
dismasted and badly
damaged *Victory* is
towed into Rosio
Bay, Gibraltar, by the
98-gun *Neptune*, after
the great storm. From
a painting by William
Clarkson Stanfield,
1854. *(City of London/
TopFoto)*

The Baltic and Peninsular War campaigns, 1808–15

After repairing at Chatham at a cost of £9,936, the *Victory* was eventually recommissioned in March 1808 as flagship to Admiral Sir James Samaurez. Her new role was to serve in the Baltic campaign, supporting the Swedes against Russia, then allied to France. Withdrawn during the Baltic winter months, the *Victory* was sent to Spain to evacuate the survivors of General Sir John Moore's army stranded at Corunna on 23 January 1809. In April the ship returned to the Baltic, blockading the ports of Kronstad and Karlskrona. Serving as flagship of Admiral Sir Joseph Yorke in 1811, the *Victory* had been temporarily converted into a troopship. Her role was to transport military reinforcements to Lisbon to support General Sir Arthur Wellesley's (Duke of Wellington's) army engaged in the Peninsular War aiding the Spanish to drive Napoleon and the French out of Spain. After further service as Samaurez's flagship in the Baltic, the *Victory* finally returned

to Portsmouth on 4 December 1812 and paid off 16 days later.

The *Victory* after 1812

Between 1814 and 1816 the *Victory* underwent a large repair at Portsmouth Dockyard during which she was very much altered and rebuilt. The ornate square beakhead bulkhead was replaced with a more practical round bow introduced by the innovative Surveyor of the Navy Sir Robert Seppings. Many of her wooden knees were replaced with a combination of wooden beam-end chocks with integral heavy iron-plate knees. Her bulwarks were raised and built up square and solid, giving her an austere appearance. As with all other ships, her sides were now painted with alternating black and white horizontal stripes and her stern decoration much reduced. With the long war against France finally at an end, the *Victory* was placed in 'ordinary' (and laid up in reserve). In 1824 the *Victory* took on a new role and was converted to act as a flagship for the Port Admiral of Portsmouth and served as tender to the larger and more modern first rate ship the Duke of Wellington. In 1831 the *Victory* was placed on the disposal list but the First Sea Admiral Hardy (her former commander at Trafalgar), at his wife's request, refused to sign the disposal order. The *Victory* had had another reprieve. Public opinion would also save her from the breaker's yard a few years later. Refitted again in the autumn of 1888, the *Victory* was coppered for the fifteenth and last time. Following this the ship became the flagship of the Commander-in-Chief and remains so today.

Disaster struck in 1903 when the *Victory* was accidentally rammed by the iron warship *Neptune* when the tow broke as she was being taken away to the breaker. (Ironically, this ship bore the same name as the ship that towed the *Victory* into Gibraltar after Trafalgar.) After an emergency docking in Portsmouth Dockyard, *Victory* went back to her familiar mooring in Portsmouth harbour and took part in the centennial anniversary of Trafalgar in 1905. This event and the previous damage incurred by the accident raised questions about her vulnerability.

In 1910 the Society for Nautical Research was founded 'to encourage research into matters relating to seafaring and shipbuilding in all ages among the nations, into the language and customs of the sea, and into other subjects of nautical interest'. It was also at this time that the Royal Navy began to take an academic interest in understanding maritime history. Unfortunately, the First World War intervened and prevented a serious review of the *Victory*'s future. Finally, through the efforts of the Society for Nautical Research, a national appeal was initiated to save the *Victory*. Soon supported by the Royal Navy, the *Victory* was placed into the dry dock she currently occupies, and was refitted and restored by naval dockyard shipwrights and riggers to her Trafalgar appearance – a living monument to the men of Nelson's Navy, moreover the Royal Navy.

First opened to the public in 1928, the *Victory* has received some 6.5 million visitors. During this time she has endured the ravages of war when damaged by a bomb in 1941, infestation with death-watch beetle and dry and wet rot.

Chapter Three

Anatomy of the *Victory*

It took timber from some 6,000 oak and elm trees to build the hull of a ship like the *Victory*. Pine, spruce, fir and beech were also used. Other materials needed in vast quantities included iron nails and spikes, iron and copper bolts and roves (or washers).

OPPOSITE Detail of lower shroud deadeyes, their lanyards and respective stopper. Deadeyes were originally made from elm or *lignum vitae*. (Jonathan Falconer)

The decks and internal arrangements

The hold

Starting in the bottom of the *Victory*'s hull is the hold, which although it comprises one compartment measuring some 51,500cu ft in volume, is divided into two parts: the main hold and the after hold. The former extends forward from the main mast and pump well to the fore bulkhead. The after hold terminates at the aftermost bulkhead. Together these holds serve as the main provision store for the ship and can contain up to six months' supplies. To counterbalance the weight of guns and masts above, the bottom of the hold is lined with ballast. If storing for six months at sea, you will need to line the bottom of the hold with 257 tons of pig-iron ballast, over which you will need to lay some 200 tons of shingle on which to bed down the lower tier of water casks.

At the centre of the hold is the pump well, which comprises a large box-like structure built to protect the casings of the bilge bumps and the elm-tree pumps. The well also contains the step for the lower part of the main mast. Integrally fitted afore and abaft the well are lidded shot lockers, each containing 40 tons of round shot for the guns. The heavy weight of this iron shot acts as ballast.

Beyond the fore bulkhead of the main hold are three hanging storerooms. That to larboard (port) is the Boatswain's block room, which also contains tar and tallow. That to starboard is the Carpenter's lower store of paint, linseed oil and pitch. On the centre line between these storerooms is the coal hole containing 50 tons of coal together with wood kindling to fuel the galley and other heating appliances. Below the coal hole is the fore shot locker containing a further 40 tons of round and miscellaneous types of bar and chain shot.

The hanging storerooms are closed off on their fore side with a thick transverse bulkhead beyond which is the grand magazine and its integral compartments. Beyond the after hold bulkhead are a series of storerooms comprising (forward to aft) the fish room, the spirit room, the bread room and a tiny compartment called the 'lady's hole', which can serve as a store for the gun-room mess.

ABOVE AND RIGHT
Pump well housing, main mast step, and pump casings of chain elm tree pumps. The author is holding the elm-tree pump casing.
(Jonathan Falconer)

RIGHT Iron and shingle ballast with shingle ballast blocks. (Jonathan Falconer)

FAR RIGHT The coal hole in the hanging storeroom over the foremost shot locker. (Jonathan Falconer)

LEFT Detail of lath and plaster and deal lining of the after magazine bulkhead, dividing off flammable stores in the hanging storerooms abaft (i.e. the coal hole and pitch room). (Jonathan Falconer)

The grand magazine

This is the main gunpowder store within the ship with a capacity to hold some 35 tons of gunpowder contained in 784 barrels of powder. This quantity of gunpowder has the explosive capacity of 47 tons of TNT and if ignited will cause considerable damage over a distance of 3 miles (5km). Because of this potential hazard the entire construction of the grand magazine had to be highly complex to avoid three dangers:

1. Explosion. To prevent explosion the entire construction is fastened together with copper nails and bolts to avoid sparks. The wing bulkheads (walls) of the pallating flat and filling room are lined with copper sheathing. The main purpose of the copper sheathing was to stop rats gnawing their way into the magazine and leaving dangerous trails of powder around the ship. The deck in the filling room is covered with thick lead. All fastenings, nails and bolts, etc., as well as hinges and door bolts, used in the magazine complex are made from copper or brass.

ABOVE The filling room in the grand magazine. Note the lead-lined deck and powder bin. (Jonathan Falconer)

LEFT Entry scuttle to grand magazine and filling room. Note lead-lined deck and copper and brass fittings. (Peter Goodwin)

LEFT Moisture-absorbent charcoal beneath the pallets that form the pallating flat. (Jonathan Falconer)

HMS *Victory* hull cutaway drawing (Colin Mudie)

HMS *Victory* – built to the design of Sir Thomas Slade, her keel was laid down at the Old Single Dock, Chatham, on 23 July 1759, and she was launched on 7 May 1765.

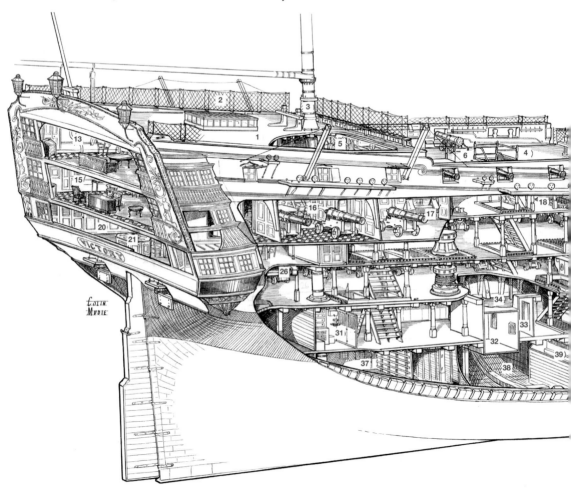

2. Fire. To prevent fire the main after bulkhead of the pallating flat is lined on both sides with layers of thin wooden boards covered with a thick layer of fire-retardant mortar and plaster laid up on oak laths. This construction system is also used for the wing bulkheads of the barrel rooms.

3. Damp. To keep the gunpowder dry all bulkheads and the underside of the deck planking of the orlop above is lined with two layers of thin wooden boards, each ¾in (2cm) thick. To prevent damp rising from the bilges below, the deck of the pallating flat is constructed in two layers, comprising four

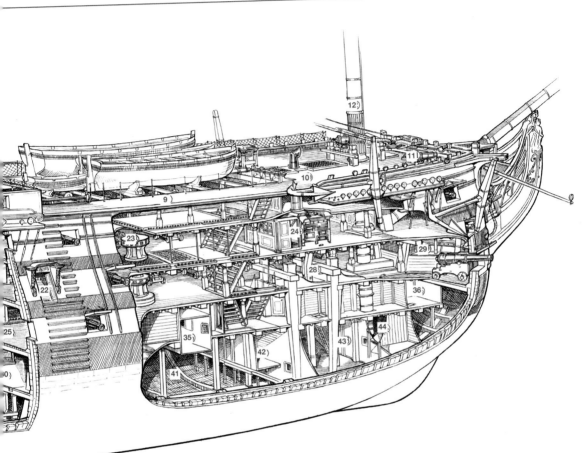

Key

1 Poop
2 Hammock nettings
3 Mizzen mast
4 Quarterdeck
5 Steering wheels
6 Here Nelson fell
7 Pikes
8 Mainmast
9 Gangway

10 Forecastle
11 Carronades
12 Foremast
13 Captain Hardy's cabin
14 Upper deck
15 Nelson's day cabin
16 Nelson's dining cabin
17 Nelson's sleeping cabin
   with cot
18 Shot garlands

19 Middle deck
20 Wardroom
21 Tiller head
22 Entry port
23 Capstan head
24 Galley and stove
25 Lower deck
26 Tiller
27 Chain and elm-tree
   pumps

28 Mooring bitts
29 Manger
30 Orlop
31 Sick bay
32 Aft hanging magazine
33 Lamp room
34 Midshipmans' berth –
   here Nelson died
35 Forward hanging
   magazine

36 Powder store
37 Powder room
38 Aft hold
39 Shot locker
40 Well
41 Main hold
42 Cabin store
43 Main magazine
44 Filling room

transverse beams 9in x 9in (23cm x 23cm)
square in cross section supporting a caulked
deck 3in (7.6cm) thick. Over this planking are
more transverse beams 7½in x 7½in (19cm x
19cm) square in cross-section beams with short
pieces between forming a grid, which subdivide
the area into square compartments covered

with portable wooden pallets 1½in thick. The
vertical void spaces formed under the pallets
are filled with moisture-absorbent charcoal.

Access to the main part of the grand
magazine complex is made via a single
hatchway and ladder leading down into the
filling room from the orlop platform above, via

ABOVE Cartridge
racks with drawer
underneath to collect
loose powder.
(Jonathan Falconer)

ABOVE Gunpowder
barrels on the pallating
flat in the grand
magazine.
Note leather hides
between barrel tiers
to prevent chafing.
(Jonathan Falconer)

containing 100lb of gunpowder (45kg). All
gunpowder barrels are banded with copper and
hazel hoops to prevent sparks. When stored,
each tier is layered with leather hides between
to prevent the barrel hoops chafing. Smaller
barrels containing either 50lb or 25lb (22.5kg
or 11.25kg) of powder are also available for
guns mounted in the ship's boats. Entry into
the pallating flat is made from the filling room by
removing the horizontal slats amidships.

The filling room: this is where the gunpowder
is converted into cartridges. Located on the
centre line is a large lead-lined oak bin into
which gunpowder is poured from the barrels for
making up the powder charges. As an instant
fire precaution there is a large lead pipe emitting
through the deck above the bin to flood the
bin with water quickly from a lead-lined cistern
fitted in the gunner's store on the deck above.
Gunpowder is poured into the lead-lined bin
and measured out into flannel or light canvas
cartridge bags using copper scoops and
measures. Paper cartridges are also used,
hence the term 'cartridge paper'. The ready use
cartridges are marked accordingly with a colour
code and stored on special racks located
either side of the filling room. These racks have
drawers underneath to collect hazardous loose
powder. Cartridges for the 12-and 24-pounder
guns are transferred to the respective hanging
magazines. Opposite the entry hatchway is a
small scuttle through which empty cartridge
pass boxes called 'cases of wood' are passed
down for refilling in battle. In battle you will
need six men working in the filling room: the
yeoman of the powder room to measure out
the charges, four landsmen to store and handle

a network of protective passageways and fire
doors. The magazine is divided into three main
sections, comprising the pallating flat, filling
room, light room and light boxes. Added to this
are subsidiary wing compartments for storing
spent gunpowder barrels.

The pallating flat: this serves as the main
powder store with up to 784 barrels, each

Entry to magazines

The keys of all magazines are to be held by the ship's Captain who
will issue them to the ship's Gunner on his request when:

1. Magazines are opened in preparation for battle.
2. When the Gunner requires to store gunpowder.
3. When the Gunner requires to have his yeoman make up
   cartridges.
4. When the Gunner requires to transfer cartridges to the hanging
   magazines.
5. When the Gunner requires to unload and transfer gunpowder
   ashore.
6 When the Gunner requires to transfer shaken down gunpowder
   barrels to the barrel rooms.

ABOVE The glazed light box behind a glazed sash with protective copper grill and wooden shutters.
(Peter Goodwin)

ABOVE Light box sashes and cover lids in the filling room.
(Jonathan Falconer)

charges and the ship's cooper to open and 'shake down' (dismantle) the barrels. Once filled, the 'cases of wood' are passed up through the main access hatchway to a chain of men supplying the 32-pounder guns. The sides of the filling room are lined with copper to prevent sparks and keep out rats. The deck of the filling room is lined with lead sheeting. Like the pallating flat, the filling room deckhead is

also double lined. At the centre line of the filling room is a complex construction comprising two segregated glazed light boxes housing the lanthorns used to illuminate the magazine.

Light boxes and light room: to avoid explosion no naked lights are allowed in the magazine. Consequently, the pallating flat and filling room are illuminated by two large

Movement of gunpowder

1. When making up cartridges or transferring cartridges to the hanging magazines this work is to be undertaken at the start of the First Watch (8pm to 12pm) when all men off watch are turned into their hammocks and all lanthorns within the ship have been checked extinguished (i.e., 'lights out') by the Master-at-Arms.
2. In battle all lead-lined passages to magazine and light room are to be filled with 2in of water to damp down loose powder.
3. In battle all cartridges are to be passed by hand in a chain of men, etc. formed in

lines from magazines to the gun decks using cylindrical lidded pass boxes called 'cases of wood', the empty cases for refilling being passed back in the reverse route.
4. According to ship's standing orders introduced c.1801, no boys are to be stationed in or near the magazines. The colloquial term 'powder monkey' given for a boy carrying powder is in fact a misconceived expression wrongly used by fiction writers, and which has probably been confused with the designated 'powder man' attached to each gun and looking after the powder box for cartridges (salt box).

lanthorns. These are located in two light boxes lined with fire-protective tin and fitted behind heavy glass sashes protected with copper grilles and wooden shutters. Brightness is increased by angled splay-boards faced with tin to reflect light. The lanthorns in the light boxes are lit in safety from the segregated light room located forward of the filling room. The light room can only be entered via a hatch sited at the end of a separate light-room passage on the orlop above. During battle, the ship's cook and the Master-at-Arms are to man this room.

The barrel rooms

All empty gunpowder barrels are eventually to be returned to the Ordnance Board for reuse; however, as they are impregnated with gunpowder they remain hazardous for the duration. Once 'shaken down' (dismantled), the staves, headers and hoops are to be bundled together and stored in the fireproof barrel rooms located in the larboard and starboard wings of the pallating flat and filling room. As a fire precaution, these barrel rooms are double lined with mortar and timber.

The orlop

Above the hold is a mezzanine called an orlop, which being multipurpose in its use is perhaps the most complex of decks in the ship. The name orlop originates from the Germanic word '*oberlaup*', or 'overlap', which means 'to overrun the hold with planks'. The centre section of the orlop is fitted with short loose boards let down between deck beams, creating a practical facility to provide complete access at any point to the stores contained in the hold below. The foremost and aftermost ends of the orlop, however, do comprise fixed platforms covering the magazine and all flammable storerooms below. The centre-line compartments on the fore platform comprise the Gunner's storeroom, the passage to the grand magazine and a passage to the grand magazine light room. The decks of these passages are lined with lead sheeting turned up 4in to 6in 'so as to hold water' when transferring powder. On the larboard side is the Boatswain's storeroom and adjacent sail room for the smaller sails, jibs and staysails, etc. Abaft this is the Boatswain's cabin. On the

starboard side is the Carpenter's storeroom with his adjacent cabin abaft. Each of these cabins and passages are entered from a small lobby called the fore cockpit, which has hatchways giving access to the fore part of the main hold and the coal hole below. Fitted abaft the fore cockpit is the fore hanging magazine with its integral light room and light box. This magazine is copper lined throughout and fitted with racks for the ready use cartridges for the 24-pounder guns. Next aft is the main sail room for the large square sails. It is fitted with wide doors at the after end to provide easy access. The open areas formed between the sail room and the wing passages at the ship's sides are the larboard and starboard cable tiers in which the anchor cables are housed. Next amidships is the main hatchway to the hold and the mast room enclosing the various pump casings. This room provides access down into the pump well. Next abaft is the after hanging magazine with its integral light room and light box. Like its counterpart forward, it is copper lined throughout and fitted with racks for the ready use cartridges for the 12-pounder guns. From this point extends the after platform.

The compartments fitted along the larboard side of the after platform comprise a cabin, for the Captain's Clerk, the marines' clothing store, the slop room and the Purser's cabin and the Ship's Stewards' room with a door leading down into the bread room. It is from this room that bread (ship's biscuit), cheese, meat, peas, oatmeal, etc. is issued. Ranged along the starboard side are the Lieutenants' storeroom and the Captain's storeroom. Farther aft is the Surgeon's cabin with its adjacent dispensary. Located amidships is the mizzen mast and a passage leading to the spirit room and bread-room light room. This was later used to provide access to the after powder room, fitted in 1808. The centre-line area enclosed by all the aforesaid compartments and storerooms is called the after cockpit. Hatchways on the centre line of the after cockpit provide access down into the fish room and spirit room. The entire orlop is surrounded by wing passages running around the ship's side. Their function is to provide free access for the ship's Carpenter and his crew to plug any shot holes sustained below the waterline in battle.

TOP A cable storage reel in the Boatswain's storeroom on the orlop fore platform.

ABOVE Carpenter's storeroom.

LEFT Boatswain's cabin – orlop fore cockpit larboard.
(All pictures this page Jonathan Falconer)

RIGHT **After cockpit starboard. From left to right: Lieutenant's storeroom, Captain's storeroom, Surgeon's cabin, Surgeon's dispensary.**
(All pictures this spread Jonathan Falconer)

BELOW RIGHT **Hanging wine cask in the Captain's storeroom.**

ABOVE **Starboard wing passage along the ship's side, colloquially called the 'carpenter's walk', which was used to carry out hull repairs below the waterline. Note the access grating to the wing store and barrel rooms below.**

The lower gun deck

This deck carries a battery of 30 long 32-pounder guns and is consequently clear of any transverse constriction from forward to aft. Along the centre line of this deck are a series of hatchways providing access down to the orlop and the hold below, each fitted with gratings to permit ventilation.

The main features ranged throughout this deck from forward to aft comprise the following. The manger, an area closed with a transverse bulkhead 4ft high fitted to prevent seawater that may enter the hawse

holes running the full length of this deck. Next is the foremast, two pairs of riding bitts and two cartridge scuttles. Further aft is the trundlehead (lower part) of the jeer capstan. Amidships surrounding the main mast are two pairs of chain pumps and one elm-tree pump. Next aft is the trundlehead (lower part) of the main capstan and the mizzen mast. This entire deck serves as the living quarters for some 480 seamen sleeping in hammocks and messing at tables set between the guns or in the space behind them. In the *Victory* hammocks are set 16in apart. Divided off with

canvas screens at the after end of this deck is the gun room. This serves as a separate mess for warrant officers, junior lieutenants and marine officers. It is furnished with two canvas cabins each side. The most senior warrant officer to live in here is the Gunner, who berths in the aftermost cabin on the starboard side. The aftermost cabin on the starboard side is to be given over to the ship's Chaplain, who is also to act as the *Victory*'s schoolmaster. Besides the various officers, the gun-room mess is equally to be shared at mealtimes by the junior midshipmen and the first-class volunteers. Each of these boys is of sufficient social status to be sent to sea to become a naval officer. All of these 'youngsters' are to be kept under the watchful eye of the Gunner. Between mealtimes the gun room is to be used as a schoolroom. The small hatches in the gun room lead to the bread room and to the 'lady's hole'. In the gun room is a large horizontal wooden lever called a tiller, which, by means of ropes and pulley blocks, transfers the helm movement put on the ship's steering wheel to the ship's rudder. Because the tiller sweeps in an arc under the overhead beams no hammocks or cots can be slung from the beams except within the canvas cabins at the ship's side.

LEFT Cupboards in the Captain's storeroom.

BELOW The four chain pumps and starboard elm-tree pump situated amidships on the lower gun deck.

Middle gun deck

This deck carries a battery of 28 long 24-pounder guns and, with the exception of access to the officers' accommodation aft, it is virtually clear of any transverse constriction from forward to aft. Along the centre line of this deck are a series of hatchways providing access down to the lower gun deck below, each fitted with gratings to permit ventilation.

The main features ranged along this deck from forward to aft comprise the following. The bowsprit step, the ship's galley with its iron stove, the galley pantry with its cupboards, the drumhead (upper part) of the jeer capstan, main mast, and the drumhead (upper part) of the main capstan. The majority of this deck serves as the accommodation for seamen and all of the marines, each sleeping in hammocks and messing at tables between the guns or in the spaces behind them. Next is a transverse, wooden-panelled portable bulkhead that divides off the officers' wardroom from the rest of the deck. The wardroom is fitted with cabins along the ship's side for the naval lieutenants and the Captain of Marines. The aftermost cabin on the starboard side is to be allocated for the ship's First Lieutenant. A pantry is fitted in the fore part of the wardroom adjacent to the mizzen mast. Doors either side lead off to the larboard and starboard quarter galleries, which serve as toilet facilities. That to starboard is to be allocated to the First Lieutenant only. At the aftermost part of the wardroom is the head of the ship's rudder,

ABOVE Tiller rope tensioning tackle in the gun room.

RIGHT Steering rope at the fore of the tiller with its gooseneck and overhead the tiller quadrant.

RIGHT A sea service long 24-pounder carriage gun in firing position, larboard battery, middle gun deck.

BELOW Marines' berth and hammocks looking aft on the middle gun deck.

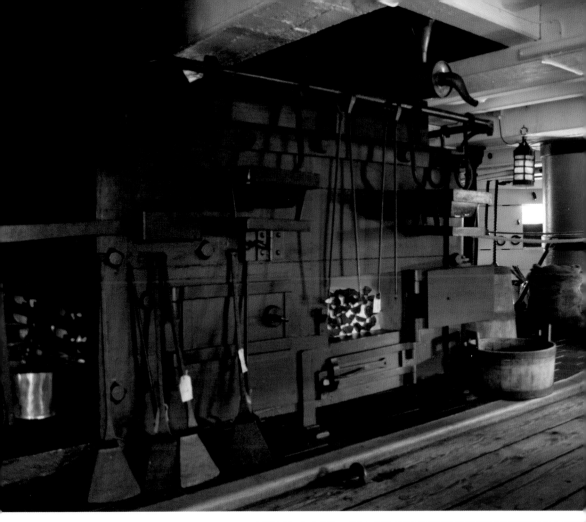

ABOVE The Brodie patent iron galley stove – starboard view showing fire box under kettles, baking oven door and fire hearth, and tools for tending the fire.

RIGHT Galley pantry and cupboards.

(All pictures this spread Jonathan Falconer)

ABOVE Detail of galley hanging stoves.

ABOVE Foremast and partners, middle gun deck.

RIGHT The mainmast at upper gun deck level.

(All pictures this spread Jonathan Falconer)

which is furnished with a horizontal mortise to receive a secondary tiller used as an emergency steering aid.

Upper gun deck

This deck carries a battery of 30 long 12-pounder guns and, with the exception of access to the Admiral's accommodation right aft, this deck is virtually clear of any transverse constriction from forward to aft. Along the centre line of this deck are a series of hatchways providing access down to the middle gun deck below, each fitted with gratings to permit ventilation. The main features ranged along this deck from forward to aft comprise the following. The larboard and starboard round-houses, which are fitted into the transverse beakhead bulkhead. These enclosures, entered by a door on their after side, serve as the heads (toilets) for midshipmen, warrant and petty officers. Next is the sick berth with its integral dispensary and its transverse and longitudinal

canvas portable bulkheads. Enveloped within this area are the foremast and a wooden steam trunk for conveying steam from the galley below up to the forecastle. Next aft are the galley flue and the galley skylight.

Amidships is the main mast and adjacent elm-tree pump. With the exception of the

ABOVE Marine's cot marked up with his ship's muster book number.

LEFT Forecastle grating over the galley skylight.

FAR LEFT Bowsprit heel and step from larboard on the middle gun deck.

RIGHT The main transverse bulkhead dividing off the Admiral's quarters from the rest of the upper gun deck. The entry door to the left leads to the Admiral's bed place; the door to the right to the steerage.

ABOVE The Admiral's dining cabin looking through to his day cabin beyond.

RIGHT Doors through the ship's side to larboard quarter gallery, Admiral's day cabin.

FAR RIGHT Seat of ease (toilet) in the larboard quarter gallery.

(All pictures this spread Jonathan Falconer)

loblolly boys tending to the sick, few people live on this deck as the midship part from the break of the forecastle and quarterdeck above is open to the elements. Divided off aft by a transverse portable panelled bulkhead are the Admiral's quarters, the foremost section being called the steerage, which is so called as earlier sailing ships were steered from this position by means of a vertical lever called a whipstaff. The steerage generally serves as an ante-room for the Admiral's valets, servants and clerks waiting in attendance. Through this compartment the tiller ropes are passed leading from the ship's steering wheel on the quarter deck above down to the tiller in the gun room.

At the after end of the steerage is the mizzen mast. Divided off to starboard of the steerage by a longitudinal bulkhead is the Admiral's bed place, containing his hanging cot slung between two guns, a washstand and other furniture. Aft of the bed place and the steerage is another transverse portable panelled bulkhead, which divides off the Admiral's great cabin proper. The foremost section of the great cabin comprises the Admiral's dining cabin with two guns either side. Here also is a companionway leading up to the ship's steering wheel on the quarter deck. Beyond a second transverse portable panelled bulkhead is the Admiral's day cabin, used for both relaxation and private secretarial functions. Doors either side lead off to the quarter galleries which serve as toilet facilities.

TOP LEFT Hanging cot in the Admiral's bed place, with replicas of drapes made by Emma Lady Hamilton.

TOP RIGHT Stern chase port lid behind bench seat and panels, Admiral's day cabin.

ABOVE Detail of the stern chase port lid, Admiral's day cabin.

ABOVE LEFT Nelson's wash stand.

RIGHT Door from the Admiral's bed place to the upper gun deck, starboard. *(All pictures this spread Jonathan Falconer)*

BELOW Starboard 68-pounder carronade.

ABOVE The belfry.

RIGHT Beak deck double 'seat of ease'.

LEFT Galley flue and steam gratings.

LEFT Grating giving access to the bowsprit and its jib booms and yards (colloquially named the 'marines' walk').

BELOW The bowsprit, cap and jack staff with jib boom, flying jib boom and all yards struck.

The forecastle

This short deck extends aft from the beakhead bulkhead to the ship's waist and serves both as an anchor deck and open area from which to operate the sails and rigging of the foremast and bowsprit. It also provides access to the shrouds and ratlines to ascend the foremast. On it are mounted two 64-pounder carronades, mounted on slide carriages and two medium-length 12-pounder guns. The main features ranged along this deck from forward to aft comprise the following: the foremast, the steam gratings, the head off the galley flue, the belfry in which is mounted the ship's bronze bell and the transverse breast rail running along the after edge of the deck. Centrally fitted at the fore end of the forecastle is a long grating forming a bridge that gives access onto the bowsprit. Suitable as a position to post a sentinel, it is colloquially called the 'marines' walk'.

The waist

This is an open area extending from the after end of the forecastle to the fore edge of the quarter deck, the two decks being joined by narrow gangways fitted along the larboard and starboard sides of the ship. Ladders fitted afore and abaft each gangway give access down onto the upper gun deck. The only other features are the transverse skid beams on which the ship's boats are housed.

front of which is the binnacle containing the magnetic compasses. Either side of the wheel are small cabins, that located to larboard was used by the ship's Master, that to starboard the Admiral's Secretary. Abaft the wheel is a transverse panelled bulkhead that divides off the Captain's quarters, comprising a dining room to larboard through which passes the mizzen mast and Captain's bed place to starboard. Beyond a second transverse portable panelled bulkhead is the Captain's day cabin, used for both relaxation and private secretarial functions. Doors either side lead off to the quarter galleries, which serve as toilet facilities.

The poop deck

This is a short open deck above the Captain's quarters and primarily serves as a conning position and signal deck for communicating flag signals to the fleet. It also provides access to the shrouds and ratlines to ascend the mizzen mast and the open area needed to operate the sails and rigging of the mizzen mast. The main features ranged along this deck from forward to aft comprise the poop breast rail, the mizzen mast, a skylight over the Captain's dining cabin and two signal flag lockers at the stern transom and the ensign staff from which to fly the red, white or blue squadron ensign. Today only the very familiar Royal Navy White Ensign is flown.

The quarterdeck

This deck extends from the main mast to the stern of the ship. On it are mounted 12 short-length 12-pounder guns. This deck serves as the control centre of the ship from which the ship is conned and sailed and battle directions and orders are given. It also provides access to the shrouds and ratlines to ascend the main mast and the open area needed to operate the sails and rigging of the main mast. With the exception of hatches and companionways, the main features ranged along this deck from forward to aft comprise the following. The main mast and the ship's double steering wheel, in

BELOW **Quarterdeck:
ship's wheel binnacle
and mizzen mast.**
(Peter Goodwin)

BELOW RIGHT
**Quarterdeck Short
12-pounder gun
starboard battery.**
(Peter Goodwin)

ABOVE Poop: skylight, taffrail and signal flag lockers against the transom. *(Peter Goodwin)*

RIGHT Ensign staff and its tabernacle formed by the taffrail transom knees. The ensign staff was removed when going to sea. *(Jonathan Falconer)*

The beak deck

This extremely short open deck extending forward beyond the transverse beakhead bulkhead is used for operating the sails and rigging of the bowsprit. Here also are the heads (toilets) for the common seamen and marines. Access out to this deck is made through doors set in the beakhead bulkhead.

The quarter galleries

These glazed galleries built on the quarters of the ship at the stern contain the heads (toilets) for the commissioned officers. Access to these is made through doors set in the ship's side.

ABOVE Signal flag locker. *(Jonathan Falconer)*

RIGHT Poop breast rail and nettings. *(Jonathan Falconer)*

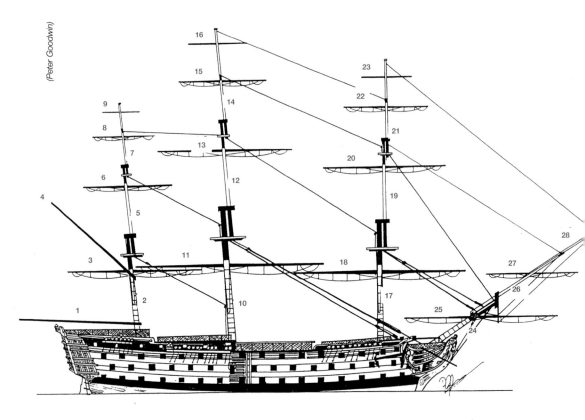

(Peter Goodwin)

Key

1 Mizzen boom
2 Mizzen mast
3 Crossjack yard
4 Mizzen gaff
5 Mizzen topmast
6 Mizzen topsail yard
7 Mizzen topgallant mast

8 Mizzen topgallant yard
9 Mizzen royal yard
10 Main mast
11 Main yard
12 Main topmast
13 Main topsail yard
14 Main topgallant mast

15 Main topgallant yard
16 Main royal yard
17 Fore mast
18 Fore yard
19 Fore topmast
20 Fore topsail yard
21 Fore topgallant mast

22 Fore topgallant yard
23 Fore royal yard
24 Bowsprit
25 Spritsail yard
26 Jib boom
27 Sprit topsail yard
28 Flying jib boom

Masts, yards and booms

The masts, yards and booms of the *Victory* are made from pine fir or spruce. Although light in weight, these species of timber are primarily used because they are rich in sap, have natural elasticity and are less likely to crack under the forces imposed upon them by the collective wind pressure upon the sails. The masts, yards and booms required for the ship comprise the following forty-three items.

Note: in the table on page 62 the dimensions have been calculated from original proportional specifications. This is because current recorded and published sources unfortunately reflect errors valued on the replacement masts put into the *Victory* in 1886, which were taken out of the ship *Shah*. Consequently, for example, the fore lower mast is too short.

RIGHT Main mast and rigging. *(iStock)*

No.	Item	Length (ft)	Maximum diameter (in)
1.	Main mast	119	37
2.	Main topmast	71.4	21
3.	Main topgallant mast (with royal mast combined)	35.7 + (25)	12 & (8)
4.	Foremast	107	33
5.	Fore topmast	64.2	19
6.	Fore topgallant mast (with royal mast combined)	32 + (22.5)	10.6 & (7)
7.	Mizzen mast	102.34	23
8.	Mizzen topmast	50	13
9.	Mizzen topgallant mast (with royal mast combined)	25 + (25)	8 & (5)
10.	Bowsprit	75.9	36
11.	Jib boom	52.5	15
12.	Flying jib boom	69.0	11.5
13.	Main course yard	107.9	26
14.	Main topsail yard	77	16
15.	Main topgallant yard	38.5	7.6
16.	Main royal yard	–	–
17.	Fore course yard	93.6	23.4
18.	Fore topsail yard	67.4	14
19.	Fore topgallant yard	33.7	6.7
20.	Fore royal yard	–	–
21.	Mizzen crossjack yard	67.4	13.8
22.	Mizzen topsail yard	48	10
23.	Mizzen topgallant yard	24	4.8
24.	Mizzen royal yard	–	–
25.	Mizzen boom	77	16
26.	Mizzen gaff	48	10
27.	Spritsail yard	67.4	14
28.	Sprit topsail yard	33.7	7
29.	Main lower stunsail boom (2)	59.44	11.8
30.	Main stunsail yard (2)	33.9	6.78
31.	Main topmast stunsail boom (2)	38.5	7.7
32.	Main topmast stunsail yard (2)	22	4.4
33.	Main topgallant stunsail boom (2)	19.5	3.85
34.	Main topgallant stunsail yard (2)	11	2.75
35.	Fore lower stunsail boom (2)	59	11.8
36.	Fore stunsail yard (2)	33.9	6.78
37.	Fore topmast stunsail boom (2)	33.7	6.74
38.	Fore topmast stunsail yard (2)	19.25	3.85
39.	Fore topgallant stunsail boom (2)	16.85	3.37
40.	Fore topgallant stunsail yard (2)	9.6	2
41.	Ensign staff	40	6.6
42.	Jack staff	20	4.9
43.	Fore course tack boomkin (2)		

RIGHT Lower shroud deadeye and lanyard.
(Jonathan Falconer)

RIGHT Forecastle rigging with medium short 12-pounder gun.
(iStock)

Rigging and blocks

Before rigging the *Victory* approximately 22,880 fathoms of hemp rope are required, which at 1 fathom equalling 6ft (2yd) in length, the overall length of cordage equates to approximately 26 miles (41.83km). This figure does not include the immense amount of spunyarn used to serve the shrouds, stays and all other requirements. The size of rope needed ranges from 19in down to ¾in in circumference.

The number of pulley blocks required for rigging the *Victory* is 768. These range from 26in treble-sheaved jeer blocks to 5in single-sheaved blocks.

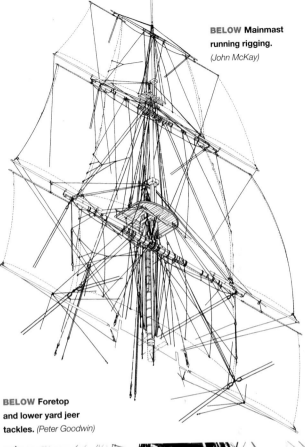

BELOW **Mainmast running rigging.** *(John McKay)*

BELOW **Foretop and lower yard jeer tackles.** *(Peter Goodwin)*

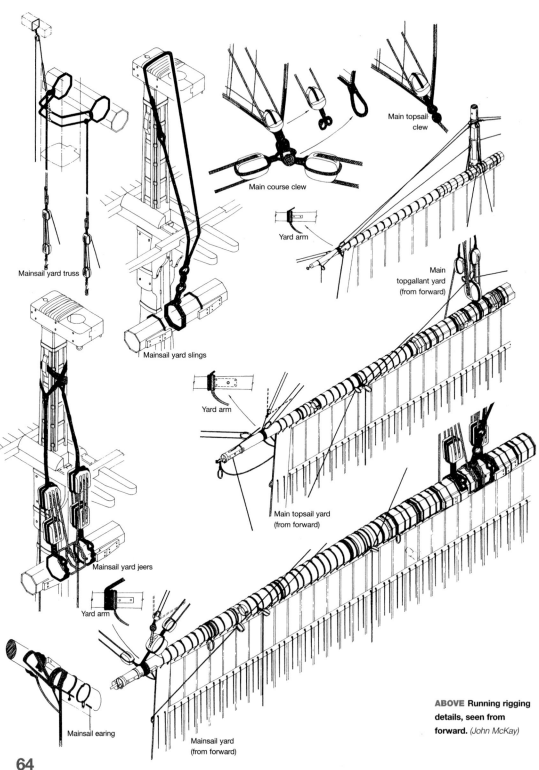

Mainsail yard truss

Mainsail yard slings

Mainsail yard jeers

Yard arm

Mainsail earing

Main course clew

Yard arm

Main topsail clew

Main topgallant yard
(from forward)

Yard arm

Main topsail yard
(from forward)

Mainsail yard
(from forward)

**ABOVE Running rigging
details, seen from
forward.** *(John McKay)*

Name of sail	Head (ft)	Foot (ft)	Leech (ft)	Luff (ft)	Depth (ft)	Area (sq ft)	Area (sq yd)
Sprit topsail	37	56	33	–	12	1,488.00	165.33
Spritsail	56	56	27	–	27	1,512.00	168.00
Fore course	88	84	46	–	46	3,956.00	439.55
Fore topsail	54	80	52	–	49	3,283.00	364.77
Fore topgallant	34	58	33	–	31	1,426.00	158.00
Fore royal	21	33	–	–	12	318.00	35.33
Main course	85	92	40	–	47	4,159.50	420.00
Main topsail	56	84	57	–	54	3,780.00	420.00
Main topgallant	36	56	37	–	35	1,435.00	159.44
Main royal	24	38	–	–	17	527.00	58.55
Mizzen topsail	40	60	45	–	42	2,100.00	233.00
Mizzen topgallant	27	41	28	–	26	884.00	98.20
Mizzen royal	15	26	–	–	10	205.00	22.70
Mizzen (driver or spanker)	55	70	69	28	–	2,618.00	290.88
Fore lower stunsail (2)	35	35	41	41	41	2,870.00	318.88
Fore topsail stunsail (2)	24	32	54	51	40	2,240.00	248.00
Fore topgallant stunsail (2)	15	23	35	32	31	1,178.00	138.88
Main lower stunsail (2)	36	36	50	50	50	3,600.00	400.00
Main topsail stunsail (2)	26	36	62	58	54	3,348.00	372.00
Main topgallant stunsail (2)	18	29	40	36	36	2,340.00	260.00
Flying jib	50	–	54	60	–	1,350.00	150.00
Jib	50	–	78	85	–	1,950.00	216.00
Fore topmast staysail	49	–	73	84	–	1,788.50	198.00
Fore staysail	43	–	46	54	–	989.00	109.88
Main staysail	57	–	43	70	–	1,225.50	136.16
Main topmast staysail	58	–	78	66	–	2,262.00	251.33
Middle staysail	48	45	39	23	–	1,395.00	155.00
Main topgallant staysail	47	40	42	20	–	1,240.00	137.77
Mizzen staysail	44	40	21	38	–	1,180.00	131.11
Mizzen topmast staysail	47	38	20	50	–	1,330.00	147.77
Mizzen topgallant staysail	39	29	8	34	–	609.00	67.00

Note: areas given for the stunsails relate to their individual sail area only, therefore multiply their areas x 2 if calculating total sail area for the ship.

Sails and sail storage

The *Victory* carries a total of 37 sails, giving her an approximate sail area of 6,510sq yd (5,468.4sq m). This area is approximately ⅓ greater than the size of a football pitch. Naturally, not all of the sails would be set at one time because wind conditions and sea state will dictate which sails are needed. For operational purposes it is best remembered that while it is a good Master who will know how much sail to set, it is an even better Master who will know

how much sail to take in. It would take 28 men 83 days to manufacture one suit of sails for the *Victory* by hand. In total, this is estimated as 64,000yd (58,240m) of seaming, each man producing 3,200yd (926m) to complete the work. Including spares, the *Victory* actually carries 59 sails.

It is extremely important that sails are not stowed away damp as damp canvas can cause spontaneous combustion. Before stowing sails away in their respective sail rooms it is recommended that they are well dried

RIGHT The steering
rope passes through
the eye fixed at
the fore end of the
tiller before passing
around the quadrant.
(Jonathan Falconer)

BELOW The head of
the rudder entering
the helm port in the
lower counter with its
tarred leather hood,
the pintles, gudgeon
braces and draught
marks. Note the
bearded fore face of
the rudder.
(Peter Goodwin)

beforehand. This can be done by easing off the
sheets and lifting them to about half their depth
using the clewlines, leechlines and buntlines
and lowering the yard on its halliard so that
they hang slack allowing the wind to dry them.
When finally unbent and sent down from their
yard they are to be carefully rolled and folded,
ensuring that their wooden tallies are visible for
identification before laying them out on the sail-
room racks.

Steering arrangements

The *Victory's* primary steering arrangements
are relatively simple, mainly comprising rope
and wood as the medium for transmission and
consequently quite prone to material failure.
The rudder, which is the main element, rotates
about its axis on bronze hinges in the form of

gudgeons strapped to the back of the stern
post and pintles fitted on the fore edge of
the rudder. The rudder can only provide a
maximum turn of about 15 degrees of helm
either side. Because the ship is propelled by
wind only and not by a power source driving
a propeller, the surface area of the rudder only
needs to be relatively small. Moreover, because
square-rigged ships tend to be steered more
by the use of sails the rudder is only a means
of providing fine adjustment. The head of the
rudder stock enters up into the ship through
a helm port cut into the lower counter of the
stern. The rudder is rotated about its axis by
a long wooden lever set in the fore and aft
plane called a tiller, which is made of ash. This
is the secondary element. The after end of the
tiller is set into a horizontal mortise cut into the
head of the rudder stock at the point of the
helm port. The foremost free end of the tiller is
supported upon a curved beam called a tiller
quadrant bolted to the underside of the middle
gun-deck beams. An iron gooseneck bolted
upon the top surface of the tiller runs freely
upon an iron flange fitted upon the upper face
of the quadrant.

Direct transmission between the ship's
steering wheel and the tiller is provided by
means of what is effectively an endless rope
rigged to eyebolts fitted at the fore end of
the tiller. This rope runs out 45 degrees from
the centre line through fixed blocks bolted at
the ship's side. It then runs forward to pass
through vertical block sheaves set in the
deckhead. These redirect the rope vertically
to the wheel on the quarter deck three decks
above where the rope is turned up around the
drum of the wheel to provide grip.

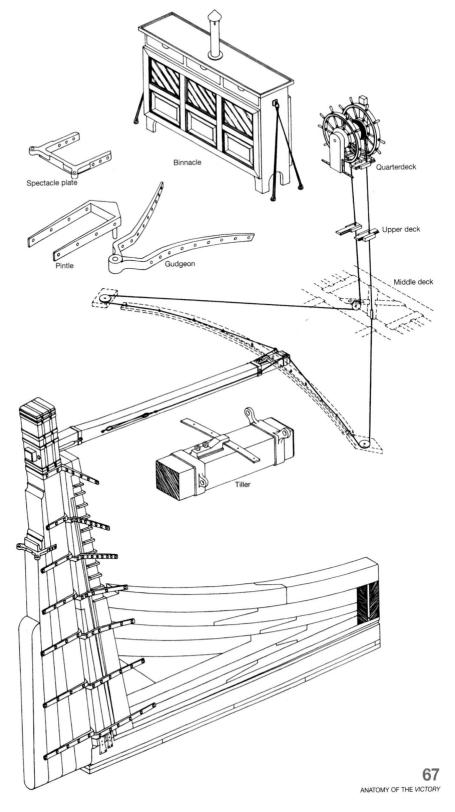

RIGHT The binnacle, just forward of the ship's wheel, contains two compasses illuminated by the lantern in the centre. The tiller sweeps in an arc in the gunroom. *(John McKay)*

Binnacle

Spectacle plate

Pintle

Gudgeon

Quarterdeck

Upper deck

Middle deck

Tiller

A copper voice tube is fitted near the ship's steering wheel on the quarter deck providing communication between the wheel and the normal helm position in the gun room on the lower gun deck below. This innovation, more associated with the destroyers and escort corvettes of the Second World War, was in fact introduced to all two- and three-decked ships as early as 1803, and the *Victory* refitting at that time would have received this modification. The introduction of this appliance proved invaluable, for when the *Victory* approached the combined Franco-Spanish fleet at the Battle of Trafalgar her steering wheel was shot away. Consequently, the ship was duly steered by use of the tiller in the gun room three decks below with helm directions being communicated through the copper voice tube. In this singular situation the tiller was operated under the direction of the *Victory*'s Master Thomas Atkinson with 10 men each side hauling on blocks and tackle rigged to the eyebolts at the fore end of the tiller.

Steering failure

The various modes of steering failure are:

- Loss of the tiller rope.
- Loss of the wheel.
- Loss of the primary tiller.
- Loss of the secondary tiller.
- Loss of the secondary steering rope.
- Damage to the rudder head.
- Loss of the rudder.

The Victory's anchors

Type	Tons	Cwt	Qtr	Lb	Tonnes
Best bower anchor	4	9	1	14	4.54
Small bower anchor	4	8	2	22	4.51
Sheet anchor (2 in number)	4	–	3	12	4.31
Stream anchor	1	1	3	7	1.10
Large kedge anchor	–	10	0	7	0.51
Small kedge anchor	–	5	3	7	0.30

The loss of either the tiller rope or steering wheel is problematic and although a new tiller rope can be rove, the *Victory*, as proven at Trafalgar, can be steered by tiller using blocks and tackle. Alternatively, the loss of the primary tiller, i.e., shot through in battle, does present initial difficulties. Therefore, set up the secondary tiller into a mortise cut into the head of the rudder stock at deck level on the middle gun deck and rerun the tiller rope or rig blocks and tackle accordingly. If both primary and secondary tillers are lost then the secondary steering ropes are to be used in unison with the steering wheel. These are external ropes permanently rigged to larboard and starboard terminating with their free ends lashed in the mizzen chains. Their lower ends pass down through eyebolts fitted on the outer extremities of the upper counter from where the ropes pass under the quarters to rudder chains linked to an eyebolt plate on the hance of the rudder. Steering can be maintained by either joining the two free ends of the two ropes and turning them up on the drum of the steering wheel or by simply applying manpower to haul upon the free ends as required. A loss of the secondary steering ropes can simply be resolved by rigging new. If the rudder head is severely damaged or completely lost whereby both tillers cannot be employed, a jury steering rig is required. For this, set up a spare yard across the quarter deck with its two ends passing well out beyond the bulwarks. Rig single blocks at the extremities of the yard through which reeve

temporary steering lines passing down to the eyebolt plate on the hance of the rudder. The loose ends of the temporary steering ropes can either be turned up on the drum of the steering wheel or operated by hand.

Loss of the rudder

According to original instructions the first thing to be done on losing a rudder, is to:

> *Bring the ship to the wind by bracing up the after yards. Meet her with the head yards, as she comes to. Take in sail forward and aft, and keep her hove-to by her sails. A vessel may be made to steer by herself for a long time, by carefully trimming the yards and slacking up the jib sheets or the spanker sheet a little, as may be required.*

Having got the ship by the wind, get up a hawser, middle it, and take a slack clove-hitch at the centre. Get up a cable, reeve its end through this hitch, and pay the cable out over the taffrail. Having paid out about fifty fathoms, jam the hitch and rack it well, so that it cannot slip; pay out on the cable until the hitch takes the water; then lash the cable to the centre of the taffrail; lash a spare spar under it across the stern, with a block well secured at each end, through which reeve ends of the hawser, one on each quarter, and reeve them again through blocks at the sides, abreast of the wheel. By this, a ship may be steered until a temporary rudder can be constructed.

BELOW LEFT
Starboard sheet anchor.
(Jonathan Falconer)

BELOW Starboard best bower anchor.
(Jonathan Falconer)

A rudder may be fitted by taking a spare topmast, or other large spar, and cutting it flat in the form of a stern-post. Bore holes at proper distances in that part which is to be the fore part of the preventer or additional stern-post; then take the thickest plank on board, and make it as near as possible into the form of a rudder; bore holes at proper distances in the fore part of it and in the after part of the preventer stern-post, to correspond with each other, and reeve rope grommets through those holes in the rudder and after part of the stern-post, for the rudder to play upon. Through the preventer stern-post, reeve guys, and at the fore part of them fix tackles, and then put the machine overboard. When it is in a proper position, or in a line with the ship's stern-post, lash the upper part of the preventer post to the upper part of the ship's stern-post; then hook

tackles at or near the main chains, and bowse taut on the guys to confine it to the lower part of the preventer stern-post. Having holes bored through the preventer and proper stern-post, run an iron bolt through both, (taking care not to touch the rudder,) which will prevent the false stern-post from rising or falling. Using the guys on the after part of the rudder and tackles affixed to them, the ship may be steered, taking care to bowse taut the tackles on the preventer stern-post, to keep it close to the proper stern-post.

Anchors and ground tackle

For operation you will need to carry the following range of anchors. Although not wholly critical, the sizes given are taken from original sources.

The best bower anchor serves as one of the two main anchors for holding the ship in deep water, and being the heaviest and strongest anchor it is to be rigged to the starboard side of the ship when operating in the northern hemisphere. Although not as strong, the small bower anchor serves the same purpose but is to be rigged to larboard. The two sheet anchors serve as spares for the bower anchors should their cables part in heavy weather or if those anchors are lost. The stream anchor is a lightweight anchor employed when anchoring in low-tide streams and shallow waters. It can also be employed for warping the ship either up river, into harbour or towards deeper water. This is done by carrying out the anchor with its cable in the ship's launch to a desired location. After turning up the stream cable upon the

The *Victory* carries 16 cables of various sizes and use:			
Type	No. carried	Circumference (in)	Diameter (in)
Cable	6	24	7.6
Cable	1	23	7.25
Cable	1	16	3
Cable	1	9	3
Cable	1	8	2.5
Cable	1	7	2.25
Messenger	1	16	5
Messenger	1	14	4.5

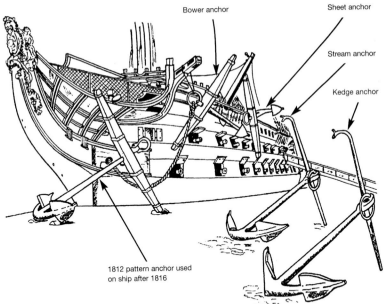

Bower anchor

Sheet anchor

Stream anchor

Kedge anchor

1812 pattern anchor used
on ship after 1816

**ABOVE Larboard
anchor cable and
messenger passing
over the manger
boards. Contrary to
popular belief it is very
unlikely that animals
were ever kept in
the manger in the
Victory, but under the
forecastle.** *(Jonathan
Falconer)*

LEFT Victory's anchors.
(Peter Goodwin)

capstan the capstan can be used to haul the
ship up to the anchor. This performance is
repeated as often as desired to place the ship
in its required position. Lighter in weight, the
two kedge anchors can be used to keep the
ship steady and clear of her bower anchor
when riding at anchor in a confined harbour,
and being relatively small they are ideal to

'kedge', or warp, the ship as described above.
Equally, kedge anchors can be employed when
laying up the ship on springs (small cables)
in order to turn the ship to align its broadside
upon a target when anchored ahead and
astern. When laying out an anchor a buoy is
always to be attached to the anchor by a buoy
rope of sufficient length.

Anchor handling

The enclosed area right in the bows is the manger, with low bulkheads to contain the muck and water scrubbed from the cable as it comes aboard. In the drawing the starboard anchor is being worked and the cable can be seen nipped to the traveller. *(John McKay)*

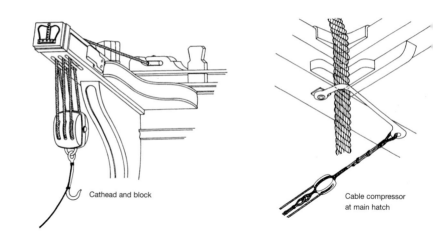

Cathead and block

Cable compressor at main hatch

BELOW OPPOSITE
Capstan bars rigged to the drumhead section of the jeer capstan on the middle gun deck.
(Jonathan Falconer)

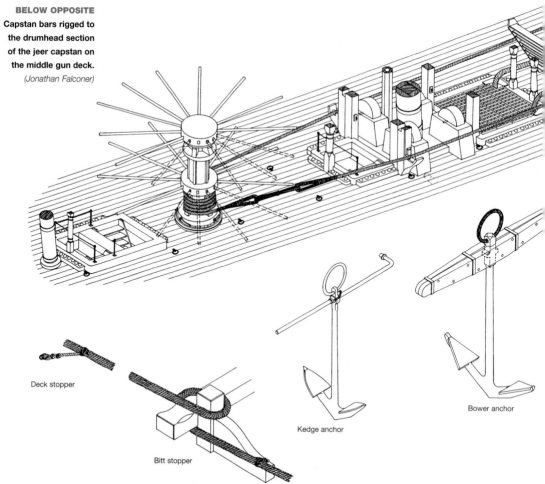

Deck stopper

Bitt stopper

Kedge anchor

Bower anchor

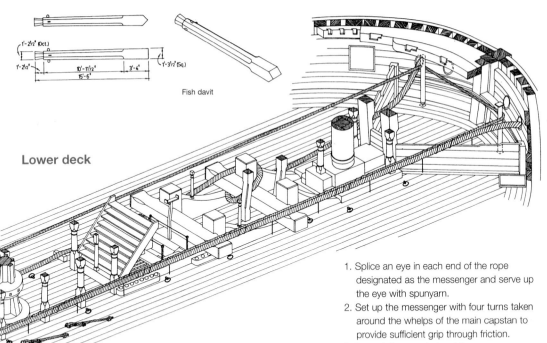

Fish davit

1'-2½" (Oct.)

1'-2½"

10'-11½"   3'-4"

15'-6"

1'-3½" (Sq.)

Lower deck

The two larger-sized cables are too great in their diameter to pass around the whelps of the capstan, therefore a smaller cable called a messenger (sometimes called a vyol) is to be employed. To enable this facility to be used the following points are to be observed:

1. Splice an eye in each end of the rope designated as the messenger and serve up the eye with spunyarn.
2. Set up the messenger with four turns taken around the whelps of the main capstan to provide sufficient grip through friction.
3. Lay out the remainder of the messenger along the gun deck and pass it through the vertical rollers located in the manger.
4. Join the two eyes at the end of the messenger together with a long lashing creating a continuous rope that passes twice around the length of the gun deck. The messenger is now ready for use.

ABOVE The outer
end of the capstan
bar with swifter.

*(All pictures this spread
Jonathan Falconer)*

7. Ease cable compressor tackle and move
   compressor clear.
8. Rig leather hoses from the lower gun deck
   elm-tree pump to the manger.
9. Prime the elm-tree pump.

Procedure:

1. Haul up sufficient slack rope from the cable
   tier and take off the turns of the anchor cable
   from the riding bitts.
2. Unlash cable stoppers and prepare.
3. Using the nippers marry (bring the anchor
   cable and messenger parallel to each
   other) and nip the two ropes together with
   a temporary lashing called a 'nipping'. This
   work is done by the nimble younger seamen
   and boys, hence the nickname 'nipper' given
   to a youngster.
4. With the main capstan manned on one or
   both decks commence turning the capstan
   whereby the messenger drags in the anchor
   cable.
5. As the anchor cable enters through the
   hawse hole you are to hose down the cable
   within the manger to remove any mud
   and weed, the residue being taken away
   overboard through large scuppers.
6. When a portion of the anchor cable
   approaches the main
   hatchway leading to
   the orlop and cable
   tiers remove the
   nippers to free the
   messenger from the
   anchor cable. The
   messenger itself will
   continue moving along
   and around the deck.

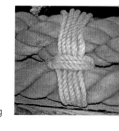

7. The young 'nippers' are to then run back
   with their nipping rope to the start point near
   the manger to take another turn marrying the
   cable and messenger together and repeat
   the process.
8. The anchor cable is to be hauled down
   through the main hatch and coiled down
   neatly within its cable tier.

Weighing anchor and bringing in the anchor cable

Prerequisites between decks:

1. Short length ropes called nippers are
   prepared and brought to the fore end of
   lower gun deck.
2. All centre-line ladders and pillars on the
   lower and middle gun decks in way of the
   turning circle of the main capstan bars are
   unshipped and taken down.
3. Take down the centre section of the
   wardroom bulkhead in the way of the turning
   circle of the main capstan bars.
4. Ensure all guns on the lower and middle gun
   decks in the way of the turning circle of the
   main capstan bars are removed or turned
   fore and aft against the ship's side and
   secured.
5. Unship the manger boards if required.
6. Unship the main hatchway gratings on the
   lower gun deck.

RIGHT Nipping
or 'marrying' the
anchor cable to the
messenger.

Maximum manpower required

Task	No.
Manning the capstan bars of the main (centre-line) capstan on both levels	260
Nippers	20
Scrubbing the hosing cable	4
Operating the elm-tree pump and tending hoses	4
Stowing the cable in the cable tier	40
Boatswain's Mates (Petty Officers) overseeing work	6
Lieutenants and midshipmen	2
Estimated total	**336**

Note: throughout this procedure the capstan is
to be turned continuously to maintain continuity
and momentum, hence the need for the young
'nippers' to work with speed and dexterity.

While the anchor cable is being hove in, make preparations on the forecastle to 'cat', 'fish' and stow the anchor.

Catting, fishing and stowing the anchor for sea

Prerequisites on the forecastle:

1. Check cat blocks are free to run, and run out the cat-block fall (rope) through its fixed sheave on the bulwark and run it aft along the gangway and make it fast to a deck cleat.
2. Check the cat-back tackles are rigged and free.
3. Set up and rig the fish davit with its heel upon the fore channel together with its topping lift and guys.
4. Rig the fish block to the outboard end of the fish davit.
5. Reeve the fish line and hook through the fish block and rig a painter (hand rope) to the hook.
6. Prepare the anchor ring strop.

Procedure:

1. When the anchor is 'up and down' (or 'a'cock bill') at the hawse hole, lower the cat block as required.
2. Using the cat-back tackles pass the hook of the cat block through the anchor ring. (It may be necessary to lower a man with the cat block to assist engaging the hook or do this from the ship's launch.)
3. Hauling in on the cat-block falls hoist the anchor to hang vertically from the cat head; the anchor is now said to be 'catted'.
4. Pass the anchor ring strop through the anchor ring and hitch it over the cat head to take the strain off the cat-block tackle.
5. Lower the fish line and engage the fish hook around the shank of the anchor in the angle formed by the arm (this action may need the assistance of the man previously lowered or men in the launch); the anchor is now said to be 'fished'.
6. Hauling in on the fish-line fall, bowse in the anchor and engage the palm of the anchor into the block fitted on the ship's side.
7. Secure the anchor for sea using the chain and stopper bolted to the ship's side.

LEFT The messenger cable is passed four turns around the whelps of the main centre-line capstan.

ABOVE On the lower gun deck the 24in circumference anchor cable is turned up on the riding (mooring) bitts. The smaller rope running along the deck is the messenger.

LEFT Fore riding bitt cable and messenger.

Ship's boats

The *Victory* carries six boats, and their type and use is as follows:

1. 34ft launch: large general working boat for conveying stores, provisions, guns, gunpowder, personnel and marines. If required, this boat can be armed with an 18-pounder carronade mounted in its bow to provide firepower for supporting amphibious operations.

2. 28ft barge: used for conveying the Admiral, high-ranking officers and dignitaries ashore or to other ships in his squadron.

3. 32ft pinnace: mainly used for conveying the officers and other personnel ashore or to other ships in the squadron. Being relatively large, it can also be used for amphibious operations.

4. 25ft cutter: sea boat and general working boat for carrying stores and provisions, also for conveying officers and other personnel ashore or to other ships in the squadron. It can also be used for amphibious operations.

5. 25ft yawl or cutter: sea boat and general working boat for carrying stores and provisions, also for conveying officers and other personnel ashore or to other ships in the squadron. It can also be used for amphibious operations.

6. 18ft cutter: small sea boat mainly used for conveying officers and their baggage and other personnel ashore or to other ships in the squadron. This is an ideal boat for transferring despatches between ships.

BELOW Admiral's 32ft barge and 28ft pinnace stowed on the skid beams. Note sheaved timber heads of breastrail and red painted arms chest for battle. *(Peter Goodwin)*

BELOW Deck plan showing locations of ship's boats. *(John McKay)*

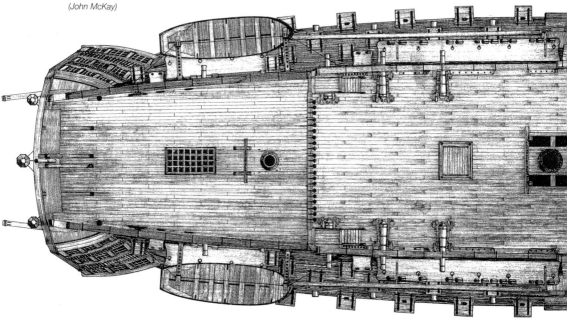

ABOVE The 34ft launch for provisioning, storing ship and amphibious attack. On the dock side are bored-out 24-pounders cast in 1846–48 on land-pattern carriages. *(Peter Goodwin)*

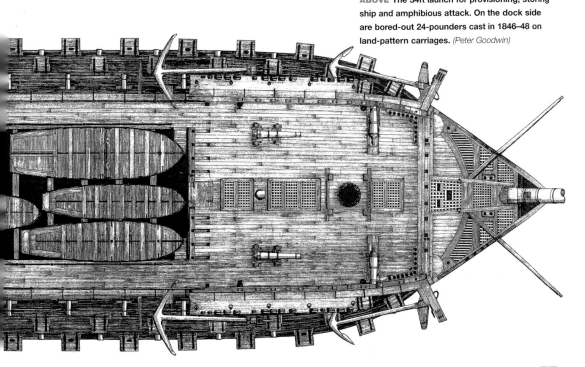

Chapter Four

Victory's Guns

At the Battle of Trafalgar the *Victory*'s formidable firepower totalled 104 guns mounted on three full gun decks, her forecastle and the quarterdeck. Each of her gun crews were trained to fire off a single round every 90 seconds, although some could achieve a round a minute.

OPPOSITE Starboard side lower and middle gun deck gun ports. *(Jonathan Falconer)*

ABOVE **HMS** *Surprise*
(right) squares off
against the better
armed and manned
French frigate *Acheron*
in this 'still' from
the film 'Master and
Commander', for
which the author was
an adviser to 20th
Century Fox. *(Topham
Picturepoint/20th Century
Fox Film Corporation and
Universal Studios and
Miramax Film Corp)*

Firepower

The weight of iron discharged from single-shotted guns fired from one broadside of the *Victory* is 1,148lb (522kg), i.e., 0.65 Imperial tons. If firing both broadsides at the same time, the combined broadside weight of the *Victory*'s firepower is doubled to 2,296lb (1.25 tons of iron). This weight is 35% greater than the massed firepower of 1,704lb (0.76 tons) discharged from the 161 guns that Field Marshal Wellington had at his disposal to support his allied army in the field at the Battle of Waterloo against the French in 1815. *Victory*'s first opening broadside at Trafalgar, fired through the stern of the French Flagship *La Bucenture*, was treble shotted, thus the weight of broadside in this instance was 3,444lb (1,566 kg), i.e., 1.9 Imperial tons.

When serving as Vice Admiral Lord Nelson's flagship at the momentous Battle of Trafalgar on 21 October 1805 the *Victory*'s ordnance totalled 104 guns mounted on 3 full gun decks, her forecastle and the quarter deck. In all, *Victory*'s armament comprised the following:

Deck	No. of guns	Gun type
Lower gun deck	30	Short 32-pounder carriage guns
Middle gun deck	28	Medium 24-pounder carriage guns
Upper gun deck	30	Long 12-pounder carriage guns
Quarter deck	12	Short 12-pounder carriage guns
Forecastle	2	Medium 12-pounder carriage guns
Forecastle	2	68-pounder carronades on slide carriages

Note: according to the Gunner's records, the *Victory* was also carrying an 18-pounder carronade, which was probably reserved for mounting in the ship's launch to use should a boat attack be called for.

42-pounder short cut (carried when first commissioned in 1778)

Length	9ft 6in	Bore diameter	6.90in
Calibre	16.244	Powder charge standard weight	14lb
Weight of gun	65cwt 0qtr 0lb	Range maximum at 6 degrees	2,740yd
Weight of gun (lb)	7,280lb	Range point blank (gun level)	400yd
Weight of carriage	13cwt 0qtr 0lb	No. of gun's crew – firing single firing	
Total weight of piece and carriage	78cwt 0qtr 0lb	on one broadside of ship only	16
Total weight of piece and carriage (lb)	8,736lb	No. of gun's crew – firing both	
Proportional weight of shot to gun	173	broadsides of ship simultaneously	8
Weight of shot	42lb	Total forces exerted upon breeching	
Shot diameter	6.68in	rope and ship's side when gun fired	18 tons

Short 32-pounder

Length	9ft 6in	Bore diameter	6.35in
Calibre	17.725	Powder charge standard weight	10.66lb
Weight of gun	55cwt 0qtr 0lb	Range maximum at 6 degrees	2,640yd
Weight of gun (lb)	7,280lb	Range point blank (gun level)	400yd
Weight of carriage	10cwt 2qtr 0lb	No. of gun's crew – firing single firing	
Total weight of piece and carriage	65cwt 2qtr 0lb	on one broadside of ship only	14
Total weight of piece and carriage (lb)	8,736lb	No. of gun's crew – firing both	
Proportional weight of shot to gun	192.5	broadsides of ship simultaneously	7
Weight of shot	32lb	Total forces exerted upon breeching	
Shot diameter	6.10in	rope and ship's side when gun fired	16 tons

ABOVE A 32-pounder gun rigged with breeching rope and preventer breeching rope. All 32-and 24-pounder guns were rigged in this manner due to their heavy recoil. With a standard charge and rope restraints this could be up to 11ft.
(Peter Goodwin)

LEFT Starboard battery of 32-pounders looking aft.
(Jonathan Falconer)

Medium 24-pounder

Length	9ft 6in
Calibre	19.574
Weight of gun	49cwt 2qtr 1lb
Weight of gun (lb)	5,545lb
Weight of carriage	9cwt 2qtr 0lb
Total weight of piece and carriage	59cwt 0qtr 1lb
Total weight of piece and carriage (lb)	6,609lb
Proportional weight of shot to gun	231
Weight of shot	24lb
Shot diameter	5.54in
Bore diameter of gun	5.74in
Powder charge standard weight	8lb
Range maximum at 6 degrees	1,980yd
Range point blank (gun level)	400yd
No. of gun's crew – firing single firing on one broadside of ship only	12
No. of gun's crew – firing both broadsides of ship simultaneously	6
Total forces exerted upon breeching rope and ship's side when gun fired	14 tons

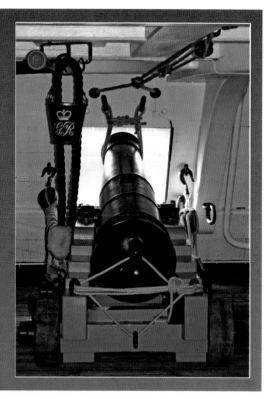

RIGHT A 24-pounder on the middle gun deck, with muzzle lashing stowed for sea. Note that the ship's bulwark has been whitewashed to improve the available light. *(Jonathan Falconer)*

Long 12-pounder

Length	9ft 0in	Total weight of piece and carriage	38cwt 1qtr 17lb
Calibre	23.361	Total weight of piece and carriage (lb)	4,301lb
Weight of gun	32cwt 0qtr 0lb	Proportional weight of shot to gun	298.6
Weight of gun (lb)	3,584lb	Weight of shot	12lb
Weight of carriage	6cwt 1qtr 17lb	Shot diameter	4.40in
		Bore diameter of gun	4.64in
		Powder charge standard weight	4.0lb
		Range maximum at 6 degrees	1,320yd
		Range point blank (gun level)	375yd
		No. of gun's crew – firing single firing on one broadside of ship only	10
		No. of gun's crew – firing both broadsides of ship simultaneously	5
		Total forces exerted upon breeching rope and ship's side when gun fired	10 tons

LEFT Starboard battery long 12-pounder on the upper gun deck. *(Jonathan Falconer)*

Medium 12-pounder

Length	8ft 6in
Calibre	22.063
Weight of gun	31cwt 2qtr 0lb
Weight of gun (lb)	3,528lb
Weight of carriage	6cwt 1qtr 6lb
Total weight of piece and carriage	37cwt 3qtr 6lb
Total weight of piece and carriage (lb)	4,234lb
Proportional weight of shot to gun	294
Weight of shot	12lb
Shot diameter	4.40in
Bore diameter of gun	4.64in
Powder charge standard weight	4lb
Range maximum at 6 degrees	1,320yd
Range point blank (gun level)	375yd
No. of gun's crew – firing single firing on one broadside of ship only	10
No. of gun's crew – firing both broadsides of ship simultaneously	5
Total forces exerted upon breeching rope and ship's side when gun fired	10 tons

Short 18-pounder carronade

Length	2ft 4in
Calibre	5.447
Weight of gun	38cwt 1qtr 25lb
Weight of gun (lb)	949lb
Proportional weight of shot to gun	52.74
Weight of shot	18lb
Shot diameter	5.04in
Bore diameter of gun	5.16in
Standard powder charge weight at 1/12	1lb 8oz
Lowest powder charge weight at 1/16	1lb 2oz
Highest powder charge weight at 1/8	2lb 4oz
Range maximum at 5 degrees using a 2lb charge	1,000yd
Range point blank (gun level) using a 2lb charge	270yd
No. of gun's crew	5

Short 12-pounder

Length	7ft 6in
Calibre	19.468
Weight	29cwt 1qtr 0lb
Weight (lb)	3,276lb
Weight of carriage	5cwt 3qtr 12lb
Total weight of piece and carriage	35cwt 0qtr 12lb
Total weight of piece and carriage (lb)	3, 932lb
Proportional weight of shot to gun	273
Weight of shot	12lb
Shot diameter	4.40in
Bore diameter of gun	4.64in
Powder charge standard weight	4lb
Range maximum at 6 degrees	1,320yd
Range point blank (gun level)	375yd
No. of gun's crew – firing single firing on one broadside of ship only	10
No. of gun's crew – firing both broadsides of ship simultaneously	5
Total forces exerted upon breeching rope and ship's side when gun fired	10 tons

ABOVE The larboard battery of quarter-deck short 12-pounders on their correct sea service carriages, beds and quoins. Note the kevel cleat between them.

(Peter Goodwin)

Short 68-pounder carronade

Length	5ft 2in
Calibre	7.702
Weight of gun	36cwt 0qtr 0lb
Weight of gun (lb)	4,302lb
Proportional weight of shot to gun	59.25
Weight of shot	168lb
Shot diameter	7.5in
Bore diameter of gun	7.75in
Standard powder charge weight at $\frac{1}{12}$	5.5lb
Lowest powder charge weight at $\frac{1}{16}$	4.25lb
Highest powder charge weight at $\frac{1}{8}$	8.5lb
Range maximum at 5 degrees using $\frac{1}{12}$ charge	1,280yd
Range point blank (gun level) using $\frac{1}{12}$ charge	450yd
No. of gun's crew	6

ABOVE This larboard 68-pounder carronade was one of two such short-barrelled weapons mounted on the forecastle of the *Victory*. Although of limited range when compared to longer guns, it was nevertheless capable of causing devastation on the decks of enemy ships with a hail of round shot and grapeshot. The *Victory* was the only ship in service to carry carronades of this heavy calibre. *(Jonathan Falconer)*

The velocity of the *Victory*'s guns

The larger 32-pounder guns are very powerful. Using a full charge of about 11lb (5kg) of gunpowder, i.e., $\frac{1}{3}$ the weight of the shot being fired, the muzzle velocity (MV) of these guns firing a single 32lb (14.4kg) round shot is between 1,500ft and 1,600ft per second (487m/sec). Because these velocities equate to between 1,023mph and 1,091mph (between 1,646m/h and 1,755km/h), consequently the projectiles fired from smooth-bored muzzle-loading guns are supersonic, the shot travelling 1 mile in 3.3 seconds. At point-blank range of 370yd (338.33m) a 32lb shot travelling at these velocities passes through 3ft (0.9m) of oak or 6ft (1.83m) of fir or pine. The MV of the smaller calibre 12- and 24-pounder guns is only marginally slower than that of the 32-pounders, albeit the firepower is proportionally reduced.

In comparison to the standard long guns, the carronades have a considerably lower MV. However, firing a proportionally heavier shot at

RIGHT The hind truck of a medium 12-pounder carriage. *(Jonathan Falconer)*

FAR RIGHT The fore truck of a medium 12-pounder carriage. *(Jonathan Falconer)*

RIGHT Detail on the solid elm truck of a 24-pounder gun carriage. The inscribed letters signify: 24 P = 24-pounder; 18 F = 18-inch diameter fore truck. The hind truck would read 24 P 16 H. *(Peter Goodwin)*

ABOVE FAR RIGHT Detail of the gun trunnion, cap square mounting and forelock retaining pins. *(Jonathan Falconer)*

RIGHT Gun equipment. From left to right: swab bucket, match tub with linstocks and slow match, and red ready-use cartridge box (colloquially called a salt box). *(Jonathan Falconer)*

a lower velocity, these guns have the advantage of creating greater damage at short range as the shot will 'churn' rather than punch its way through timber. Attempting to effect a watertight hull repair to this kind of damage is far more difficult than stopping up a clean hole produced by standard guns.

Manufactured in cast iron, all of the standard guns in the *Victory* are of the Blomefield pattern introduced in 1787 to supersede the Armstrong pattern, which, often bursting at the breech, proved a failure during the War of American Independence. Not only did bursting guns cause unnecessary fatalities to gun crews, gunners became reluctant to fire these guns with a full charge. Blomefield guns were designed with more metal reinforcing the breech and were less likely to burst subsequently. This modification gave the British gunners greater confidence when firing their ordnance and consequently they attained superiority in gunnery throughout the French Revolutionary and Napoleonic wars. Following advances in the technology of gun manufacturing the vent and the bore of the Blomefield guns were bored out after casting. This process produced more accurate guns.

Side arms and gunnery equipment

Each gun was to be furnished with the following tools, termed 'side arms':

- Rammer – to ram home shot, charge and wads.
- Sponge – to swab out the gun with water to extinguish smouldering debris.
- Wad hook (or worm) – to remove wads and burning debris.
- Combined rammer and sponge.
- Combined flexible rope rammer and sponge – used when loading or swabbing with the gun-port lids shut.
- Two wooden handspikes made of ash – to manoeuvre the gun carriage sideways when pointing or traversing the gun.
- Ladle – used for inserting a cartridge.
- Cartridge pricker – thin brass pointed skewer 10in long.

ABOVE A realistic display of shot damage to a ship's internal timbers, created by the author with the assistance of a 32-pound gun from Royal Armouries Fort Nelson and volunteers of the National Museum of the Royal Navy. *(Peter Goodwin)*

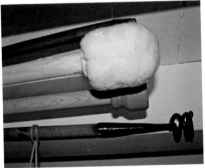

LEFT Gun side arms. From front to back: sheepskin sponge, rammer and wad hook. *(Jonathan Falconer)*

LEFT Spring-operated gunlock with flint and lanyard. *(Jonathan Falconer)*

ABOVE A linstock and slow match were used if the gunlock failed. *(Jonathan Falconer)*

- Vent reamer – steel or brass twisted skewer to clear vent of carbon deposits.

Items 1, 2, 3, 4 and 7 are fitted on wooden staves made of ash. Although the ladle could be used for inserting a cartridge, it was more generally used to unload a cartridge and shot on misfire.

RIGHT Muzzle tompion and lanyard. When the gun was not in use the charge, wad and shot already prepared inside was protected from damp by inserting a tompion. *(Jonathan Falconer)*

ABOVE The protective lead aprons used to cover the gunlock, touch hole and vent of the gun. Note the breaching rope lashing to 'bowse' in the slack rope, when securing the gun run-out. *(Jonathan Falconer)*

RIGHT Powder horn, leather fire bucket and combined flexible rammer and sponge. *(Jonathan Falconer)*

Other necessary equipment included:

- Wooden swab bucket.
- Wooden sand scuttle.
- Wooden match tub (two types: open top or closed top).
- Slow match: short lengths of specially manufactured rope soaked in potassium nitrate.
- Wooden linstock to hold the slow match.
- Wooden 'salt box' to contain ready use charges (lidded powder box with leather hinges and brass fittings).
- Iron crows – crow bars used as handspikes.
- Gunlock – spring-operated device that mechanically strikes a flint to produce an ignition spark.
- Firing tubes – preformed goose quills filled with very fine gunpowder mixed with spirits of wine.

Firing the guns

The *Victory*'s gun crews were to exercise at the 'great guns' at least once weekly, although the use of live ammunition and powder might need to be restricted in order to conserve powder and ammunition. To this end, the crews were at least to go through the motions of gun drill using the implements and associated equipment as required, with individual gun crews occasionally chosen to fire a live round 'at a mark'. The objective was to train the men to fire off a single round every 90 seconds, albeit records show that some gun crews could achieve a round a minute. It is often stated that the introduction of the gunlock used simultaneously with the preformed goose-quill firing tube increased speed of gunnery. This unfortunately is a misconception, although it certainly provided accurately timed firing. The major problem with the gunlock is potential mechanical malfunction or failure of the flint, whereas the slow match (unless inadvertently extinguished) provides instant ignition. Firing tubes are equally consistent with ignition by slow match. Once at sea and clear of an anchorage, all guns are loaded with a charge, wadding and shot ready for action. These contents contained within the bore of the gun are kept dry by means of a wooden tampion stopping up the gun muzzle. To avoid accidental firing the touch hole and vent of the gun are covered with a lead apron.

Important instructions:
1. Stop the touch hole. This complies with an 18th-century Health and Safety ruling: it is highly important that the vent is 'served' by stopping it up with your thumb throughout the sponging and clearing of the gun process for the following reasons:
■ To prevent burning embers being drawn down though the vent into the bore before loading a new charge. A gunpowder charge inadvertently igniting while being loaded will kill or severely injure the loader.
■ To prevent burning embers being driven up the vent and left around the touch hole when the sponge is rammed down the bore.
■ Moreover, sponging the stopped vent creates a good vacuum that extinguishes burning embers when the sponge is withdrawn. In short, the vent must be continuously served while side-arm tools are being placed into the bore of the gun, the thumb only being removed when the cartridge and wad are hard up against the inner end of the bore of the gun.

2. Sponge the gun.
   In sponging the Gun, the Spunge is to be drawn back-wards and forwards two or three times as well as pushed home strongly, and in taking it out, turn it two or three Times in the Gun (i.e., scour the bore). Observe to strike your Spunge well on the Muzzle of the Gun to Cleanse it. If you make use of a Rope Spunge, observe to shift Ends and have your Rammer Head at Hand.

LEFT A full-scale model section of a gun breech showing the cartridge, shot and wad in the bore. *(Peter Goodwin)*

BELOW A midshipman prepares to give the order to 'Fire!' *(Topham Picturepoint/20th Century Fox Film Corporation and Universal Studios and Miramax Film Corp)*

The gun drill (authorised procedure)

Item	Words of command	Observation or action taken
1.	Take heed	To get the attention of the people
2.	Silence	To prevent the people talking
3.	Cast loose your guns	Cast off side tackles, breeching ropes and muzzle lashings
4.	Seize the breechings	Ensure ends of breeching rope well secured to the ship's side ringbolts
5.	Take out the tampion	Remove stopper from muzzle of gun
6.	Take off the apron	Remove lead apron protecting vent of the gun
7.	Unstop the touch hole	Remove wax plug from vent
8.	Handle the pricking wire	Take cartridge pricker ready in hand
9.	Prick the cartridge	Firmly pierce cartridge by inserting cartridge pricker
10.	Handle the powder horn	Take powder horn ready in hand
11.	Prime	Pour powder from powder horn into the pan and vent
12.	Bruise the priming	Firm down powder with knuckle to prevent it blowing away
13.	Secure the powder horn	Keep powder horn away from gun
14.	Take hold of the apron	Take apron ready in hand
15.	Cover the vent	Replace apron over vent to prevent inadvertent firing
16.	Handle your crows and handspikes	Take up crows and handspikes ready in hand
17.	Point the gun to the object	Using crows and handspikes move gun carriage
18.	Lay down your crows and handspikes	Lay crows and handspikes down on deck out of way
19.	Take off the apron	Remove lead apron
20.	Take your match and blow it	Take slow match ready in hand on its linstock from match tub well behind the gun
21.	Fire	Shout 'make ready' and touch off the powder in the pan and stand clear
22.	Stop the touch hole	Stop the vent with your thumb (refer to Important instructions below)
23.	Handle the sponge staff	Take sponge staff ready in hand
24.	Sponge the gun	Insert sponge into the bore of gun (refer to Important instructions below)

RELOAD

Item	Words of command	Observation or action taken
25.	Handle the cartridge	Pass cartridge from salt box to loader at muzzle of gun (refer to Important instructions below)
26.	Put it into the gun	Refer to Important instructions below
27.	Wad your cartridge	Load wad: this helps compression and bed for round shot
28.	Handle the rammer	Take rammer ready in hand
29.	Ram home wad and cartridge	Using rammer drive in cartridge and wad shot hard up against the bottom of the bore of the gun
30.	Unstop the touch hole	Stop thumbing the vent
31.	Handle the pricking wire	Take cartridge pricker ready in hand
32.	Try if the cartridge be home	Using pricker feel cartridge is well in the end of the bore
33.	Draw the rammer	That the cartridge is proved well home, withdraw rammer completely from gun
34.	Shot the gun	Load round shot
35.	Wad	Load a second wad to prevent the round shot rolling out
36.	Ram home wad and shot	Using rammer drive in wad and shot hard up against the cartridge
37.	Draw the rammer	Withdraw the rammer completely from gun
38.	Stop the touch hole	Stop the vent with your thumb.
39.	Lay on the apron	Replace apron over vent to prevent inadvertent firing
40.	Run out the gun	Take up side tackles and haul gun out through gun port

3. Put it into the gun. 'You must put the Cartridge in as far as you can reach with your Arm, the lower End first, and Seam of the Cartridge downwards.'

4. Ram home wad and cartridge. 'Observe to give it two or three strokes to ram it well home.'

5. Run out the gun.
If you exercise the Lee Guns and it blows fresh you must keep one Tackle to the Ringbolt on the Deck, near the Coamings and the other Tackle hooked to the Ring in the Train of the Carriage. But if you exercise the Windward Guns, keep both Tackles hooked to the Ship's Sides and the Train of the Carriage.

6. Fire. 'You must take care, that the Guns do not touch the Sides of the Port, when you fire.'

7. 'When you exercise the Lower Deck Guns, have your Port Ropes or Port Tackle Falls clear, to let fall your Ports in case of too much Wind, and the Laniards to make them fall fast.'

8. 'Always, after the Exercise is over, take care to have the Decks clean swabbed, that no scattered Powder be left.'

Although the authorised gun procedure outlined above is effectively that laid down to 'Exercise the Great Guns', as a realistic gun drill it proves impractical amid the noise and confusion of battle. Despite the fact that controlled broadside firing used in the opening of battle creates the best destructive effect, after initial closing with an enemy ship it is necessary for gunners to achieve a high rate of fire such as one round in less than 1½ minutes. To achieve this speed the actual

**BELOW 1 The gun has fired and recoiled, the Gun Captain checks that the 'swabber' has sponged and extinguished all burning debris from the bore before loading with cartridge.
2 Swabbed and reloaded, the crew run out the gun through the gun port ready to lay, prime and fire.
3 Gun Captain 'laying the gun' as crew member lifts the breech to adjust elevation.
4 Gun Captain inserts the firing tube into the vent, checks the gunlock flint and resets the gunlock prior to priming with powder.**
(Martin Bibbings)

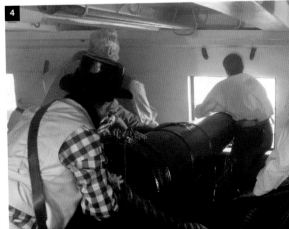

The gun drill (practical procedure)

Having already fired off one round, the simpler practical procedure for gun drill in battle would be more likely to be as follows:

Item	Words of command	Observation or action taken
1.	Stop the touch hole	Serve the vent with your thumb
2.	Search the gun	Use wad hook to clear out any debris (every four rounds)
3.	Sponge the gun	Insert wetted sponge down into bore of gun
4.	Load with cartridge	Pass cartridge from salt box to loader at muzzle of gun
5.	Load with wad	Load wad to provide bed for round shot
6.	Ram home wad and cartridge	Ram home both to compress the charge
7.	Prick the cartridge	Pierce the cartridge
8.	Shot the gun	Load round shot
9.	Wad the shot	Load and ram second wad to prevent shot rolling out
10.	Draw the rammer	Withdraw the rammer completely from the gun
11.	Stop the touch hole	Stop the vent with your thumb
12.	Run out the gun	Take up side tackles and run out the gun through gun port
13.	Point the gun	Move gun carriage as required using crows and handspikes
14.	Unstop the touch hole	Stop thumbing the vent
15.	Prime the gun	Prime the pan and bruise the priming
16.	Make ready	Take slow match ready in hand on its linstock from match tub and blow it – gun's crew stand clear
17.	Fire	Touch off the powder in the pan and stand clear

BELOW **Gun crews engage the enemy ship.** *(Topham Picturepoint/20th Century Fox Film Corporation and Universal Studios and Miramax Film Corp)*

gun drill (wholly coordinated by individual gun captains) is somewhat far simpler than the formal version, omitting such procedures that complicate and slow the process, although never compromising safety. In reality, naval gun drill was wordless as little or no verbal orders could actually have been heard above the noise of battle. If a deaf man still remains a competent gunner, the fact that he cannot hear orders would only be of consequence if there were orders to be heard. The requirement to maintain consistent high rates of fire is solely reliant on having highly trained gun crews working like automatons, oblivious to all around them. I have personally experienced this while taking part in a re-enactment as a ship's gun captain within a battery of guns using black powder where all individual gun crews simply focused on firing and reloading the gun.

Whichever gun drill is used, the introduction of the spring-operated, flint-brass gunlocks and quill-firing tubes instead of the slow match will alternate the procedure. Therefore, where the drill states 'Prime the gun', the following actions should be applied:

- Insert firing quill-firing tube into the vent.
- Remove wax cover from firing tube.
- Ease back the frizzen plate to expose the gunlock pan.
- From powder horn pour powder into gunlock pan and a small trail in vent pan to firing tube.
- Ease the frizzen plate over the gunlock pan to prevent powder being inadvertently removed.
- Check edge of flint is keen.
- Half cock the gunlock hammer.
- Run out the gun.
- Point the gun.
- Fully cock the gunlock hammer.
- Taking the gunlock trigger lanyard in hand stand to rear of gun and sight along the gun.
- Make ready – tension the trigger lanyard.
- Fire tug trigger lanyard to activate the gunlock.

The salt box containing the ready use charges brought up from the magazines is to be placed well behind the gun and constantly manned by the designated Powderman of the gun crew. His responsibility is to ensure that at least two ready use cartridges are available in the box. More importantly, to prevent an inadvertent explosion

on deck the Powderman is to ensure that the lid of the box remains shut at all times and is only opened when taking out a cartridge when the gun is being loaded. If the ship is firing broadsides on one side only, the salt box is to be placed on the opposite side of the deck to the gun being used and consequently well clear of the gun-firing process and the match tubs. If firing both broadsides simultaneously, the salt boxes for each gun are to be placed directly on the centre line, i.e., equidistant between the opposing guns.

TOP The author 'touching off' a 24-pounder gun with linstock and slow match.
(Colin Burring)

ABOVE Nets were suspended from the deck head with ready use wads for battle.
(Jonathan Falconer)

LEFT A spring-operated gunlock and trigger lanyard.
(Jonathan Falconer)

Ropes, tackles and cordage associated with guns

Breeching ropes

Contrary to popular belief, the guns do not recoil right back into the centre of the ship when used in action. Restrained by the large breeching rope, the fixed length of which was proportional to the length of the gun, the gun only recoils back some 18in to 24in (48cm to 61cm). In general proportion the breeching rope is to be three times the length of the gun (muzzle to breech ring). Although this restriction provides sufficient room to carry out worming, sponging and reloading, this limitation equally reduces the distance that the gun has to be run out again. This factor saves time, manpower energy expended and increases firing rate. When the 32-pounder gun is fired the total forces exerted upon the breeching rope and the ship's side is about 16 tons; for the 24-pounders, 14 tons; and for the 12-pounders, 10 tons. Because of their weight and recoil forces, preventer breeching ropes are also to be rigged to the 32- and 24-pounder guns. These are not necessary for the much lighter 12-pounder guns. Each gun is to be rigged with the following breeching ropes:

Side and train tackles

Each gun is to be rigged with two side tackles and one train tackle. Each comprises single- and double-sheaved blocks. The side tackles for running the gun out are to have their single blocks hooked into the eyebolts on the cheeks of the carriage; their double blocks hook into the eyebolt fastened at the ship's side. The double block of the train tackle is to hook into the eyebolt fitted into the rear axletree, the

ABOVE Breeching rope, breeching ring bolt and side tackle lashed for sea. (Jonathan Falconer)

RIGHT A long 24-pounder gun on sea service carriage, larboard battery, on the middle gun deck. Note the breeching rope and preventer breeching rope. (Jonathan Falconer)

single block hooks into the ring bolt near the centre line of the deck. Besides serving to haul the gun backward for loading, the train tackle can be used to turn the rear of the carriage to 'point' the gun obliquely. To do this, it is necessary to hook the train tackle between the eyebolt on the hind part of the cheeks and the quarter eyebolts fitted equidistant at the ship's side between each gun port. Then you have to slew the carriage round by hauling on the tackle falls aided by handspikes.

Gun-port ropes

Each gun-port lid is to be rigged with lanyards, runners and tackles (see panel below).

Muzzle lashings

When stowing guns for sea each gun is to be supplied with a length of 2in-circumference rope (16mm diameter), 4 fathoms (24ft) long to lash the muzzle up to the eyebolts fitted into the lintel of the gun port above each gun.

LEFT A 24-pounder gun with muzzle lashing, tackle lashing and breeching lashing stowed run-out.

(Jonathan Falconer)

Cordage associated with gun carriages and gun-port lids

Gun type	No.	42-pounder	32-pounder	24-pounder	12-pounder	68-pounder carronade
Breeching rope circumference	1	8in	7in	6½in	5½in	7in
Breeching rope length		30ft	28ft 6in	28ft 6in	27ft 25ft 6in 22ft 6in	15ft
Side tackle	2	3½in	3in	2½in	2in	2in
Train tackle	1	3½in	3in	2½in	2in	2in
Train tackle carronade	2	–	–	–	–	2in
Muzzle lashings	1	2in	2in	2in	1½in	–
Muzzle lashing length		24ft	24ft	24ft	24ft	–
Port ropes	2	3in	3in	–	–	–
Port tackles	1	2½in	2½in	2½in	2in	–
Port tackle runners	1	2½in	2½in	2½in	2in	–

RIGHT Starboard
broadside, with vertical
fenders accentuating
the curvature of the
ship's 'tumble home' to
the waterline.

ABOVE Lower gun
deck gun port.
(All pictures this spread
Jonathan Falconer)

RIGHT Boxes of
canister shot, a kind
of anti-personnel
munition similar to
grapeshot, but with
smaller and more
numerous balls.

Gunpowder and its properties

Gunpowder is the propellant used to fire the various types of solid iron shot or other projectiles out of smooth-bore naval guns, muskets and pistols. Rated as a relatively high explosive, it is only 35% weaker in destructive power than modern trinitrotoluene (TNT). When gunpowder is ignited it expands 300% of its own volume, providing a very effective propellant. Besides its use in naval and military gunnery and hand guns, by 1803 gunpowder was also being used for rockets both for signalling and as weapons. In the 18th century British gunpowder was made from the three materials described in the table below.

Gunpowder constituents

Chemical	Quantity	Purpose
Potassium nitrate (saltpetre)	75%	Provides oxygen to increase the combustion rate
Charcoal (carbon)	15%	Provides fuel and heat in the form of carbon dioxide to assist the combustion process
Sulphur	10%	Acts as a catalyst (chemical reaction) with the charcoal in the combustion process

In brief, the saltpetre provides extra oxygen, the potassium nitrate, sulphur and carbon react together to form nitrogen and carbon dioxide gases and potassium sulphide. The expanding gases, nitrogen and carbon dioxide, provide the propelling action.

In the manufacturing process the three ingredients are finely ground and mixed together with water to form a dough-like compound. This is then baked and milled to form granules of varying grades, the finer powder being used for quill-firing tubes, hand-grenade fuses, muskets and pistols.

LEFT 'Cases of wood' for conveying cartridges from the magazines.

FAR LEFT A box of grenades waiting to be 'fixed' (charged) with gunpowder and fuse.

LEFT Inside the Gunner's storeroom on the orlop fore-peak is a rack of India-pattern sea service muskets.

Gunners' stores

When preparing the *Victory* for sea service it is necessary to embark the following quantities of gunpowder and ball cartridges:

Common gunpowder			Copper rivet			Ball cartridges			
Whole barrels		Half-barrels	Whole barrels		Half-barrels	Issued made		To make up on board	
Channel service	Foreign service	Channel service	Channel service	Foreign service	Foreign service	Muskets	Pistols	Muskets	Pistols
431	479	5	200	–	–	7,000	2,000	10,000	4,000

Note: whole barrels contain 100lb of gunpowder and half-barrels 50lb.

When preparing the *Victory* for sea service the Gunner has to embark the following quantities of:

Ammunition, gunnery stores and equipment

Gun type size	Item or equipment	Quantity
32-pounder	Round shot	2,400 (34.3 tons)
24-pounder	Round shot	2,800 (3.0 tons)
12-pounder	Round shot	4,200 (22.5 tons)
32-pounder	Grape shot (9 x 3lb ball)	90 (1.1 tons)
24-pounder	Grape shot (9 x 2lb ball)	112 (0.9 tons)
12-pounder	Grape shot (9 x 1lb ball)	168 (0.7 tons)
32-pounder	32-pounder case shot	–
24-pounder	24-pounder case shot	–
12-pounder	12-pounder case shot	56
Boxes	For grape shot	57
68-pounder carronade	Round shot	84
68-pounder carronade	Case shot	7
68-pounder carronade	Grape in tin case	7
18-pounder carronade	Found shot	–
18-pounder carronade	Case shot	–
18-pounder carronade	Grape in tin case	–
32-pounder	Double-headed shot	90
24-pounder	Double-headed shot	84
12-pounder	Double-headed shot	126

Item or equipment		Quantity
Paper cartridge	For 32-pounder guns	2,580
Paper cartridge	For 24-pounder guns	2,996
Paper cartridge	For 12-pounder guns	4,728
Staves	Spare for ladles	20
Sheepskins	Dozen odd	50
Sponge tacks	Copper	3,000
Budge barrels	(See note 6 below)	4
Copper hoops	On ditto (budge barrels)	16
Tampions	For 32-pounder guns	120
Tampions	For 24-pounder guns	112
Tampions	For 12-pounder guns	168
Tampions	For 68-pounder carronades	6
Tampions	For 18-pounder carronades	2
Copper adzes		2
Copper driver	For copper hoops	2
Vice		2
Match	cwt	13
Lanthorns	Muscovite (mica)	3
Lanthorns	Tin	6
Lanthorns	Dark	2
Powder horns	These are additional to the 104 required	95

Item or equipment		Quantity
Priming irons		450
Aprons of lead	Large	125
Aprons of lead	Small	44
Hand screws	Large	2
Hand screws	Small	2
Handcrow levers	6ft long (handspikes)	87
Handcrow levers	5ft long (handspikes)	63
Crows of iron	5ft 6in long (crow bar)	58
Crows of iron	4ft 6in long (crow bar)	42
Ladle hooks	Pairs	20
Inch pins	Pairs for carriage trucks	10
Forelock keys	(Cotter pins)	200
Pin mauls		3
Cases of wood	For 32-pounder guns	75
Cases of wood	For 24-pounder guns	70
Cases of wood	For 12-pounder guns	105
Nails	40d (forty penny)	300
Nails	30d (thirty penny)	300
Nails	20d (twenty penny)	600
Nails	10d (ten penny)	1,200
Nails	6d (six penny)	1,500
Baskets		40
Breechings	7in	60
Breechings	6½in	42
Breechings	5½in	63
Breechings	4½in	
Breechings	3½in	
Tackles	3in (complete)	174
Tackles	2½in (complete)	126
Tackles	2in (complete)	126
Port ropes	3in	64
Port ropes	2½in	?
Port ropes	2in	
Port tackles	2in	64
Port tackle runners	2½in	60
Muzzle lashings	2in	30
Thimble straps	2in	69
Thimble straps	1½in	50
Double thimbles	For 32-pounder guns	36
Double thimbles	For 24-pounder guns	33
Double thimbles	For 12-pounder guns	50
Tarred rope	7in (coil)	1
Tarred rope	6½in (coil)	1
Tarred rope	5½in (coil)	1
Tarred rope	4½in (coil)	1

Item or equipment		Quantity
Tarred rope	3in (coil)	4
Tarred rope	2½in (coil)	1.5
Tarred rope	2in (coil)	1.5
Blocks 10in single	Pairs	14 .5
Blocks 10in double	Pairs	44.5
Blocks 8in single	Pounder	10.5
Blocks 10in double	Pairs	10.5
Blocks 6½in single	Pairs	10
Tackle hooks (pairs)	Large	50
Tallow	2cwt and 2qtr	280lb
Marline	Skeins	60
Junk (old rope)		10 tons
Hand grenades	Fixed (made up for ready use)	200
Boxes for hand grenades	20 per box	10
Copper measures	Powder for 32-pounder guns	2
Copper measures	Powder for 24-pounder guns	2
7-barrelled guns		20
Black muskets	Sea service (see notes 1 and 2 below)	150
Slings for muskets		150
Bayonets		150
Bayonet scabbards		150
Bayonet frogs		200
Musketoons		2
Pistols	Pairs (see note 3 below)	70
Cartouche boxes		200
Belts for cartouche boxes		200
Boxes of cartridge	For muskets	4
Boxes of cartridge	For pistols	1
Musket rods	Wood (see note 5 below)	75
Musket flints		5,400
Pistol flints		7,000
Musket shot	6cwt 0qtr and 18lb	690lb
Pistol shot	1cwt 0qtr and 5lb	117lb
Fine paper	Cartridge paper: reams and quires	4 and 2
Pole axes		60
Swords (cutlasses)		200
Sword scabbards		200
Sword belts		200
Strong pikes	Half pikes (see note 4 below)	50
Halberds		2
Drums		2
Spare drumheads		2
Funnels of plate		2

Item or equipment		Quantity
Sweet oil (or train oil)	(See notes 7 and 8 below)	11gal.
Copper funnels	For powder	9
Formers	For musket cartridge	24
Formers	For pistol cartridge	12
Measures	For musket cartridge	9
Measures	For pistol cartridge	4
Dutch thread		3lb
Twine		1lb
Large knives		6
Scissors		3
Hair broom	Long for magazine	1
Hair broom	Short for magazine	1
Cartridge boxes	Common for 32-pounders	18
Cartridge boxes	Common for 24-pounders	18
Cartridge boxes	Common for 12-pounders	12
Quill tubes	Firing tubes	
Boxes for quill tubes		?
Straps for quill tubes		300
Brass cannon locks	Gunlocks	50
Lock tools		50
Flannel cartridges	For 12-pounders	58
Cured paper	Covers for 12-pounders	1 set
Tin case shot	For 12-pounders	18
Boxes for tin case shot		18
Powder horns	Improved pattern	234
Sponges with staves	New pattern for 32-pounders	25
Sponges with staves	New pattern for 24-pounders	150
Sponges with staves	New pattern for 12-pounders	6
Sponges with rope	For 32-pounders	6
Sponges with rope	For 24-pounders	8
Sponges with rope	For 12-pounders	20
Sponges with staves	Common for 12-pounders	18
Combs for sponges		28
Spare sponge heads	For staves (for 32-pounder guns)	8
Spare sponge heads	For staves (for 24-pounder guns)	7
Spare sponge heads	For staves (for 12-pounder guns)	2
Spare rammer heads	For staves (for 32-pounder guns)	2
Spare rammer heads	For staves (for 24-pounder guns)	2
Spare rammer heads	For staves (for 12-pounder guns)	4
Spare sponge heads	For ropes (for 32-pounder guns)	5
Spare sponge heads	For ropes (for 24-pounder guns)	4
Spare sponge heads	For ropes (for 12-pounder guns)	7
Spare rammer heads	For ropes (for 32-pounder guns)	10
Spare rammer heads	For ropes (for 24-pounder guns)	8
Spare rammer heads	For staves (for 12-pounder guns)	14

Notes on the stores and small arms

1. Compared to the military land pattern musket with a barrel length of 42in the sea service musket has a barrel length of only 39in, which being shorter makes for a far more manageable weapon for use within the confines of a ship. Although having a shorter barrel reduces its effective range, long range is not so necessary in a close-action sea fight compared to the long ranges required in land warfare.

2. The comment referring to muskets being 'black' relates to the barrels being unpolished as opposed to being the 'bright' muskets used by the marines.

3. Pistols are always counted and issued in pairs.

4. Pikes: although formally listed as pikes, this term is inaccurate inasmuch as what ships actually carried were half pikes, 7ft in length. A pike in the proper sense is a military land weapon, 14ft in length, and consequently impractical for use in ships. Furthermore, pikes carried in ships have often been wrongly termed as boarding pikes. There is no such weapon and they are simply called a pike and are, in fact, used as a defensive weapon being rather too impractical for use as an offensive weapon.

5. Rammers for muskets: sea service musket ram rods are made of wood as these do not rust and jam in the bore. This also applies to sea service pistols.

6. Budge barrels: small cask (quarter powder barrel (25lb)) with copper and wooden hoops, and one head replaced with a leather hose or bag furnished with a drawstring to close it. These are used for carrying or holding loose powder (often for mortars) in safety from sparks.

7. Sweet oil: this is a mild vegetable oil that is either olive, rape or linseed based.

8. Train oil: this is manufactured from reducing whale blubber and used for the internal mechanism of gunlocks.

RIGHT An arms chest and a double-headed bar shot in the Gunner's storeroom. *(Jonathan Falconer)*

RIGHT Another arms chest, cutlasses, scabbards and match tubs. *(Jonathan Falconer)*

LEFT An original hand grenade and beech fuse recovered from the wreck site of the 74-gun ship HMS *Invincible*, which foundered at Spithead in 1758. *(Colin Burring)*

ABOVE Sea service flintlock pistols ready for battle. *(Jonathan Falconer)*

ABOVE A grappling iron, often thrown with a rope for grasping and holding an enemy ship prior to boarding. *(Jonathan Falconer)*

LEFT A box of sea service India-pattern muskets ready for battle. This weapon saw extensive use in Europe by the Army and Navy during the French Revolutionary and Napoleonic Wars. *(Jonathan Falconer)*

Chapter Five

General Maintenance and Refitting

◁━━●━━▷

The *Victory* is an organic structure. Her hull, decks, masts and yards are all made from timber; her rigging and sails are manufactured from other natural materials. Being organic these structures and components are prone to failure, so maintaining the *Victory* at sea was labour intensive and required huge stockpiles of 'spares'.

OPPOSITE Shipwrights use power tools to bore bolt holes in the lining (quick work) wrought between the deck clamp (beam shelf) and spirketting. *(MOD/Crown Copyright)*

aintaining a first rate ship at sea is not only highly labour intensive, requiring many of the crew to provide the physical manpower for the work, but also considerable stockpiles of material are consumed in the process. Wooden warships are virtually organic in nature with all of the hull, masts and yards made from timber, and the associated running and standing rigging, sails and their pulley blocks all manufactured from natural materials and fibre, which also being organic are constantly subject to failure. Ship's log books and journals contemporary to the period held at The National Archives make constant reference to material failure and the measures undertaken to effect ongoing repairs. This problem is wholly understandable because of the environment these warships operated in. For not only were there the elements of

corrosive seawater, wind and rainwater to deal with, deterioration was an inherent problem due to the continuous stresses imposed upon the rigging and sails, added to which were the forces made upon the ship's hull by the constant movement of the sea.

Hull maintenance at sea

Inherent hull movement at sea in the transverse plane, termed 'racking', also causes the fastenings of hanging and lodging knees at the beam-end knees and beam-end chocks to work loose. Again, maintenance is limited at sea, although the ship's Carpenter and his crew can re-drive the bolts in the hanging and lodging knees. In the case of the beam-end chocks with their integral iron-plate knees, tightening up the structure can be achieved by driving opposed iron wedges together which are fitted between specially formed lands (faces) between the chock and the beam itself. These are especially evident under the lower gun-deck beams, where racking had more effect at the ship's extreme breadth at the waterline. Access to these are obtained from the internal wing passages, colloquially called the 'carpenter's walks', which run continuously around the ship's side at the level of the orlop deck. Major refastening of knees and chocks will be dealt with during refit.

> **Racking** is the tendency for the ship's hull to move out of square to the transverse perpendicular and thereby beams are distorted at their extremities and stressed and flexed at their centres. Vertical pillars between decks are also misaligned.

Dockyard maintenance and refitting

While major maintenance of the wooden fabric of the hull is very much covered by periodic dockyard refits, some general hull maintenance can be done at sea. As with any wooden ship or boat, the seams of the ship's side planking above the waterline can open at sea and although more difficult to facilitate, some remedial action can be taken by stopping up

BELOW Lower gun deck beam-end chock and iron plate knee, wing passage orlop. Note the space between the chock and the ships' side to ventilate the end of the beam, and the opposed iron wedges used to tighten the beam-end chock when the ship is working or 'racking' in a heavy sea. *(Peter Goodwin)*

the seams with pitch.

Repairing open seams below the waterline presents a greater problem. This will involve re-caulking the seams of deck planking using oakum and pitch to make them watertight. Oakum is a fibre made from recycled old or worn hemp cordage mixed with Stockholm tar to form a malleable, greasy, yarn-like material that can be driven into the deck seams. This is done using a wedge-shaped caulking iron and a wooden caulking mallet made from beech furnished with copper sleeves around its head. The mallet should 'ring' when the oakum is driven fully home into the seam. On completion the seam is to be stopped up with molten pitch poured from a lipped ladle and the excess pitch scraped away to leave a smooth, clean surface. Dockyard caulkers were paid 1s (5p) for every 100ft of seaming worked (which equates to roughly £1.60 at today's values).

Bales of oakum were carried in the ship as part of the Carpenter's and Boatswain's stores. Today, all of the *Victory*'s gun decks are caulked using the traditional caulking material and methods described above. The only exceptions are the exposed 'weather decks', i.e., the forecastle, waist, quarter deck and poop. Here the low-maintenance, modern yacht compound Sikaflex is used.

Oakum is manufactured by unravelling old rope and teasing out the fibres using the fingernails. This painstaking task was a common penal occupation in prisons and workhouses during the 19th century, or was alternatively done by men constrained as punishment in the ship itself. The fibres produced are then hand rolled, perhaps on the thigh, and mixed with Stockholm tar.

Stockholm tar is exported from Scandinavia and is a viscous, sticky fluid

ABOVE Middle and upper gun deck gun ports with the main channel, its deadeyes and lower main stunsail boom. *(Jonathan Falconer)*

produced by burning the wood and roots of pine trees in specially designed kilns, which maintain a steady low temperature. As the pine burns the dense pitch runs out and is collected in a chamber located underneath the kiln. This process is called 'destructive distillation' because it requires complete destruction of the wood to extract the valuable sap inside.

Although inherent hull movement in the longitudinal plane, called 'hogging' and 'sagging', would open up the seams of the *Victory*'s external ship's side planking, making good the caulking was quite difficult at sea. Therefore, maintenance for this was limited and reliant on oakum and paint alone and very much left to dockyard refits.

Hogging is the tendency for the ship's hull to bend downward at the fore and aftermost extremities when the centre of length is sitting on the crest of a wave, thereby stretching the hull and its planking at its centre of length.

Sagging is the tendency for the ship's hull to bend upward at its centre of length when the fore and aft extremities are sitting on the crest of a wave, thereby compressing the hull and its planking at its centre of length.

Painting the ship

If the ship is moored alongside a hulk in harbour, painting her should be deferred until last thing and only then as circumstances will permit. The ship she should then be hauled off to moorings with the ship keepers left on board only. If an application is to be made, painters will be sent from the dockyard to paint her. If the painters work can be performed equally as well by the ship's company, it is better to demand the paint, and have it done by the people on board.

Great care should be taken in mixing the paint, which should be tried on a clean board previous to laying it on noting that the colour will always appear two or three shades darker than when dry. If the paint is mixed too thick

it dries too soon, peels off, and does not go so far as it would if mixed thinly. If, however, it is not properly consistent, it will have a very poor appearance and will not last long also if the quantity of oil is too much it could take a long time to dry.

The painter, or man required to mix the paint is to have sufficient judgment on the consistency and viscosity of the paint.

Spirit of turpentine is frequently mixed in to make it the paint dry quicker but as the sun always extracts the spirit, this practice often makes the work blister.

The following recipe is the simple easy and approved method of mixing the paint. Take the proportions of yellow and white paint, oil (linseed) and litharge that will make it of the intended colour and consistence.

Litharge is a form of lead oxide (PbO) and is made up of hard yellowish-red crystals which are created by heating lead in air. This compound provides a glaze effect to the paint.

Mix them all the ingredients together into one of the boatswain's fish kettles and stir them well up, and boil the composition. Pour it off, and strain it through a bread bag, and using paint brushes lay it on warm; the warmer it is laid on the better. To avoid making a mistake care should be taken to mark the lines before commencing painting.

If a streak [horizontal band along the ship's side] is to be painted, it should be done first with the upper works; the ship's sides should be well scraped, and where any grease may have lodged it should be scrubbed with warm lime water, and not a particle of dirt or dust suffered to remain.

There have been various fashions in painting ship's sides, but the following instruction appears to be the most generally approved:

A yellow streak, from the ribbands of the channels forward and aft, cutting a line through the ports; another streak on the moulding above it. The broad yellow from the lower part of the upper deck port holes to the lower

1 **Preparing the oakum to run into the seams.**

2 **Various caulking irons.**

3 & 4 **Driving the oakum into the plank seams with caulking iron and copper-ringed caulking mallet.**

5 **Driving the oakum until the caulking mallet 'rings'.**

6 **Ship restorer Tommi Nielson caulks the flat of the middle gun deck.**

(All pictures this spread Jonathan Falconer)

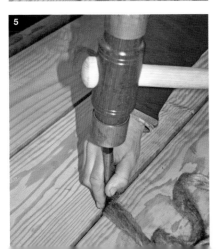

part of the quarter gallery; the lower part of the yellow to be painted in a straight line; the upper part with a small sheer, carried entirely forward to the cut water. Frigates are generally painted with a bright yellow side, and black upper works; the yellow about two inches above the upper part of the ports, and carried down to the line of the lower part of the quarter gallery. The muzzels [sic] of the guns and port-sills black. Tomkins [tampions] in the waist white, and those belonging to the guns on the quarter-deck black, to prevent the height and irregularity of the carronades from being observed.

The bends, which comprise the main wale and planking below, are to be: 'blacked with varnish and oil, mixed and boiled; otherwise, with varnish and tar. The first mixture carries the best appearance, but is not so durable or as serviceable as the latter: but the three different compositions are frequently united together, and this, perhaps, forms the best coating near the waterline.

'The middle and channel wales are to be painted with black varnish or black paint, likewise the ship's upperworks above the drift rail.'

The following general rule may also be observed: 'If it is desired to make a ship look long and low, the yellow part should be carried low down and painted narrow; if, on the contrary, it is wished to make her look lofty and short, the yellow part should be broad and carried high up.'

According to Nelson's instructions issued to his fleet in 1805, flagships were to be painted four times per annum, the other rated ships three times annually. With regard to the yellow paint used on the ship's side in 1805, Nelson instructed his own fleet that the mixture was to be 'six parts white to one part yellow ochre'.

In addition, ships were painted by contract, and unless the captain of a private ship was prepared to pay the painters for additional white to be mixed with the yellow, it was a very dark colour that made the ship look very dirty and dismal (refer to Nelson's order above). Small battens, or sinnet, were generally nailed on to point out the lines as a guidance to paint by. The stern was to be painted with black varnish or black paint with mouldings, carvings, rails, balustrades and stern light (window) munnions, etc. in yellow. Likewise, the ship's name was to be in Roman lettering at the centre of the upper counter. The quarter galleries were to be similarly painted as the stern. The beakhead bulkhead was to be painted with a blue ground with mouldings, rails and pillars, etc. painted yellow.

Gun ports and their lids

The external faces of gun-port lids are to be blacked with varnish and oil, mixed and boiled; otherwise, with black paint. The internal faces and the edges are to be red ochre.

The gun-port sills, lintels and vertical sides are to be red ochre.

Painting between decks

For practical reasons, all panelled transverse portable bulkheads contained within the Admiral's, the Captain's and wardroom quarters are to be painted a flat-finish white paint.

The exposed faces of the cabin bulkheads under the poop deck are to be painted with a protective oil-based paint. The colour, which is debatable, is to be either yellow panels with black or just plain white.

The ship's bulwarks along the exposed surfaces of the quarter deck are to be coated

BELOW Muzzles of the larboard battery of 24-pounders run out for action. Note vertical fender bottom right and the curve of the hull's 'tumblehome' toward the lower gun deck port lids below.
(Peter Goodwin)

with an oil-based yellow paint. This protective paint coating is also to be applied to the internal sides of the upper gun deck from the fore side of the transverse bulkhead dividing off the Admiral's quarters. This is because most of this deck is partially exposed to the elements.

The *Victory*'s records of 1805 show that whitewash was extensively applied to the internal sides and deckheads (underside of deck planking) throughout most of the ship, the orlop and its cabins. Whitewash was also applied to the hold deckhead. Reasons for using whitewash comprise the following:

- Brighten up the closed interior of the ship.
- Lime kills off bacteria in living areas.
- Easy to apply.
- Inexpensive.

The whitewash was made up of fish glue, sea salt, slaked lime and water mixed together and boiled in a cauldron before application.

Maintenance of masts and yards at sea

Because of the constant wind forces acting upon the sails and the rolling and pitching of the ship, the masts and yards and booms of the *Victory* are made from pine, fir or spruce, which although light in weight, produce timber that has natural elasticity. Simple maintenance comprises painting and varnishing. This also applies to the oak crosstrees and trestletrees, which provide a housing facility to support topmasts and topgallant masts at their points of juncture. The following is recommended for preservation. Lower masts and the bowsprit are to be painted a dull yellow ochre. Topmasts, topgallant masts, stunsail booms, jib boom, flying jib boom, ensign and jack staffs are to be oiled (see pages 106–8 for paint and colour schemes). All yards, mastheads, mast caps, crosstrees and trestletrees to be

ABOVE The ship's
name in Roman
lettering on the
upper counter.
(Jonathan Falconer)

to provide some way of sustaining the general rig. These comprise the fore, the mizzen crossjack and the lower spritsail yard, each of which are of approximately the same length and diameter as the fore topsail yard.

Extant log books and journals of ships continually remind us that masts sprung (cracked along their grain) during high winds and squally weather render them too weak to withstand wind pressure upon the sails. The best temporary solution at sea is to 'fish' the mast with lengths of timber lashed to the mast in the form of a splint. Typical ready use timber to employ as 'fishes' are capstan bars and stunsail booms. In the case of lower masts, it is advisable to use spare anchor stocks. Severely damaged topmasts and topgallant masts can simply be unshipped and sent down on deck and repaired or modified if possible by the Carpenter. If a lower mast breaks near the deck, its stump can be removed using jury rigged sheer legs to act as a crane. While this concept is theoretically plausible, in practice it is extremely difficult to execute unless the ship is anchored in a very flat sea. Damaged main and fore course yards may require considerable

painted black or coated in black varnish, and likewise the underside of the main, fore and mizzen tops. Although the mizzen boom, gaff and lower stunsail booms are today currently painted black, it is highly likely that these were to be oiled or alternatively treated with black varnish. As the flat of the tops is made of pine or fir boards, it is probable that the upper surface of the tops is to be left natural wood.

While the loss or damage of a yard might prove problematic to replace at sea, selected yards were designed to be interchangeable with the all-important fore topsail yard in order

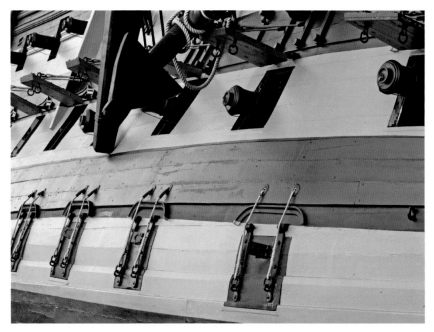

LEFT Starboard broadside with lower gun deck ports shut. *(Jonathan Falconer)*

BELOW Larboard bower anchor of 4¼ tons 'catted' ready for 'fishing'. Dock side anchors comprise an 1812 pattern bower anchor cast at Chatham and a 1-ton iron-stocked stream anchor. *(Peter Goodwin)*

of all cordage associated with the moving and hoisting of the sails, yards and booms and to some degree topmasts and topgallant masts. Complex in its application, running rigging comprises three-strand hawser laid ropes, varying in size from 1in to 10in in circumference running through 768 blocks. Although the ship carries coils of spare rope, each some 220yd long, it is impractical to carry coils covering the full range of sizes required. In fact, many repairs are simply undertaken by splicing broken ropes together. Where this is not practical, it is necessary to replace the entire length of rope required. The rope strops on the blocks or the pendants and lashings that secured the blocks in place will also need to be replaced.

Maintenance of the standing rigging is more difficult at sea mainly because its prime function is to support and brace the masts. Unlike the hawser laid rope used for running rigging, the type of cordage used for standing rigging can either be three- or four-stranded cable laid rope or shroud laid rope, the sizes varying between 2in and 18in in circumference. Furthermore, maintenance of the standing rigging is limited because it is impractical to carry spare coils of the variations of rope type and the large sizes employed.

The shrouds, providing lateral support to masts, can stretch and become slack due to the rolling motion of the ship at sea. They can be tightened up by hauling up the lanyards laced through the wooden deadeyes rigged to the lower ends of the shrouds, drawing the pair of deadeyes together. Should this not prove successful, it is necessary to use a jigger tackle. This is temporarily rigged with its upper end seized to the shroud well above the upper running deadeye and its lower end hooked to the chain of the lower fixed deadeye secured to the chain platform on the side of the ship. Using the jigger tackle, you take up the slack in the shroud then haul up tight on the deadeye lanyards to reset the shroud. It may be necessary to repeat the process until the shroud is reset. Moreover, it may also be necessary to adjust all respective supporting shrouds in the line to even out the tension. Tightening shrouds

ABOVE A 24-pounder gun on correct sea service carriage bed and quoin, rigged with breeching rope, and preventer breaching rope side tackles with handspikes ready to manoeuvre the gun. All 24-pounder guns were rigged in this manner due to their heavy recoil. *(Peter Goodwin)*

attention from the Carpenter and can be sent down and fished with timber in the same manner as a mast.

Sending down the topgallant mast at sea is a common emergency evolution undertaken to reduce top hamper aloft in sudden squalls, and can be achieved in less than half an hour. Should it be necessary to lower the topmast, this is equally achievable but does necessitate more work. In both cases, shrouds and stays can remain rigged if the mast is dropped as a temporary measure. Whichever the case, the procedure is the reverse of getting the mast up.

Maintenance of sails and rigging at sea

Because the masts' rigging sails are effectively the ship's main propulsion unit and subject to continuous use, movement and stresses, the risk of breakdown and failure is high. As a consequence, rigging maintenance at sea is constant. This is particularly predominant with the running rigging, which collectively consists supporting the lower masts presents few problems mainly because you can firmly stand on the chain platform (or channel) housing the lower fixed deadeyes. However, the task of resetting topmast shrouds at sea is, from personal experience, far more difficult mainly because you are working at the edge and below the rim of the main or respective top. Despite the implication that you are working at height above deck, the process is identical to that described above. Adjusting or fixing the topgallant mast shrouds is far more precarious, albeit you can have a foothold on the topmast ratlines or in the 5in-wide crosstrees, but nevertheless you are now working some 165ft above the waterline. Adjusting the slack out of the backstays and breast stays at sea is a relatively simple task depending on whether rigged with deadeyes or a tackle purchase at the ship's side. In short, the same rules given above apply.

The fore stays, which provide fore and aft support to masts, also become slack due to the inherent pitching movement of the ship in a heavy sea. These, like the shrouds, can be adjusted by tightening the lanyards rove between with the wooden hearts bent to the lower part of the stay or turned into the fixed rope collar secured to a point in the ship or adjacent mast. Topmast and topgallant mast stays are either rigged with deadeyes and lanyards, or alternatively the fall of the stay (lower end of the stay) is led through a thimble and secured on deck where slack can easily be taken in as required.

All standing rigging is to be regularly tarred to protect it from the weather. This process, called 'blacking down', can be done by hand using pots of pitch and brushes. The task is relatively easy to achieve in safety by ascending the lower, topmast and topgallant shrouds on their respective ratlines (footropes), allowing hands to be free. Blacking down the backstays and fore stays, however, will require considerable rope-climbing skill and dexterity as it involves precariously descending the full length of the stays with both pot and brush. Blacking down can often leave tar splashes on the masts and the deck. A vigilant Master or Boatswain should be quick to 'turn the hands to' to scrape clean offensive splashes

upon completion. Because the heads of shrouds, stays and back stays pass over the heads of respective masts, replacing them at sea will present considerable problems as it involves sending down masts, yards and their associated running rigging.

Rigging repairs sustained in action or storm damage

The practical solution and quick remedy to get the ship into a safe, operable condition relies very much on the practice of 'knotting and splicing', with ropes being replaced with new at the earliest opportunity if practicable and if not at the next dockyard refit.

Maintenance of sails at sea

By virtue of the natural fibrous material from which they are woven (flax), their relative fragility, the continuous movement either by wind or being set, stretched or furled, together with the alternating effects of dampening and drying, the sails of the *Victory* present by far the highest maintenance task undertaken in the ship. Moreover, the aforementioned conditions will promote the sailcloth to tear apart in sudden squalls or unexpectedly to rip if set wrongly in certain wind conditions or manoeuvres. As sails become defective or worn by general use, it will be necessary to replace individual sails with spares (if carried in the ship). If replaced, the ship's sailmaker and his crew should immediately begin patching and stitching the torn sails as defects appear, applying a variety of techniques common to their trade. If necessary, the sailmaker can make a new sail using bolts of sailcloth carried in the ship, albeit this stock is limited, but it is more likely he can use other old sails until replacements can be obtained from the dockyard sail lofts.

One bolt of sailcloth is 39yd long and 24in broad, the breadth limited by the size of looms available in the 18th century.

Sails are particularly prone to shot damage in battle. This is very evident from the condition and appearance of the *Victory*'s original foretopsail, which was set on its respective yard during the Battle of Trafalgar and is now on public display in the National Museum of the Royal Navy at Portsmouth. Not only does this potent and awe-inspiring object stand testament to the ravages of enemy fire the *Victory* received as she closed towards the enemy line, this foretopsail is pieced with some 90 shot holes of various sizes caused by both round, bar and grape shot. Furthermore, there is a rent some 25ft in depth running from the head of the sail caused by the foretopgallant mast as it fell when shot away in the battle.

Maintenance of the ship's boats

The Boatswain and the Carpenter are regularly to survey all boats to ensure they are seaworthy and that their hull planking is watertight. Regarding the yawl and the cutters, which are clinker (lap strake) built, it is necessary to inspect that all copper roves joining the plank edges are well driven. The remaining carvel-built boats are to be checked

that the caulking is in a good state of repair. Bottom boards are to be inspected and likewise the leather-work lining the rowlocks. This particular item will need constant replacement due to wear by the oars. Rudders are to be examined to ensure they are free from damage and that their pintles hang freely in their gudgeons. It must be checked that the bung has been removed so that the boat is drained to prevent rot occurring along the keelson. Also inspect that all oars are present and blades free from splits; all associated masts and spars are present and that their rigging and boat sail(s) are in good order; boat hook(s) and anchor are present; and the tiller is in good condition. In the case of the launch, inspect the windlass to make sure it is well greased and turns freely and its handspikes are in good order. Boats are to be painted at least annually. Excluding the Admiral's 28ft barge, all boats

BELOW The 25ft cutter hanging in its quarter davit falls with tiller lashed over and bow painter secured in the main chains.

(Peter Goodwin)

are to be painted as follows:
- External hull up to the rubbing strake to be coated in a lead-based white paint. Transom and rudder to be a lead-based white paint.
- Rubbing strake to be black paint or black varnish.
- One or two strakes below rubbing strake can be painted with black paint or black varnish. Wash boards and upper edge of gunwale to be yellow ochre paint.
- Internal hull to be yellow ochre paint or a yellow ochre pigment mixed with linseed oil. Thwarts to be of yellow ochre paint or yellow ochre pigment mixed with linseed oil or just plain linseed oil.

The paint scheme for the 28ft Admiral's barge is to be as follows:
- External hull up to the rubbing strake to be coated in a lead-based white paint.
- Transom and rudder to be a lead-based white paint.
- Rubbing strake and strake above to be verdigrease (verdigris, dull green) with mouldings picked out in gold.
- Wash boards and upper edge of gunwale to be verdigrease.
- Internal hull to be verdigrease paint or a

verdigrease pigment mixed with linseed oil.
- Thwarts to be of verdigrease paint or just plain linseed oil.
- Stern sheets to be of verdigrease paint picked out with Prussian blue or gold.

Maintaining the guns

The Gunner is periodically to inspect all guns for obvious defects and report any failings to the Captain and the Ordnance Board. The Gunner must also make arrangements for replacement as necessary. Likewise, all guns are to be thoroughly surveyed by the Inspector of the Ordnance Board. This is normally done when guns are embarked when the ship is storing for sea service. When disembarked the ship is paid off and decommissioned and all guns are returned to the nearest gun wharf of the Ordnance Board. Here they are subjected to rigorous examinations not unlike quality control employed today. These tests comprise the following:
- Examine the bore of the gun for signs of pitting and wear caused by the firing process and the effect of shot.
- Inspect the condition and diameter of the vent and ream out if required. The most

common problem found is that the bore of the vent increases in diameter by age and firing, causing considerable back fire and potential injury to the gunner firing the piece. It has become practice to line the vents of old guns with copper vent tubes tapped and screwed into the old vent. Guns thus fitted are stamped with the letters CV on the cascable ring. Replace copper vent with new as required or modify gun to take copper vent and mark gun accordingly.

■ Sound the entire casting of the piece with a hammer to detect potential cracks or defects.

■ Seal the vent and carry out a hydrostatic test (high-pressure, water-pressure test) exerted upon the bore of the gun from the muzzle. Any flaws in the form of cracking in the casting are detected by water emerging through the external surfaces of the gun.

■ Should a gun be found 'unfit for service' through either of the above examinations then smash off one or both of the gun trunnions making it impossible to mount

the gun on a carriage. Alternatively, spike the vent by hammering a large nail or bolt into the vent preventing the gun being fired. This action is reversible, albeit with some difficulty, and therefore the common practice of removing the trunnions is the best method of rendering a gun unserviceable.

Many of the old guns seen today, for example, acting as bollards in streets and at corners of dockyard buildings, are actually defective guns. The author proved this point when guns employed in this way were dug up on a development site in Portsmouth – all were found to have had their trunnions removed.

Although made of cast iron, the surfaces of guns are subjected to oxidation with weather and seawater. The various methods of protection and compounds to use are:

■ Black guns with a mixture of 6oz of lamp-black, 3 pints of spirits of turpentine and

RIGHT A 24-pounder gun hoisted off its carriage to allow repairs to be made to the hind truck.
(Peter Goodwin)

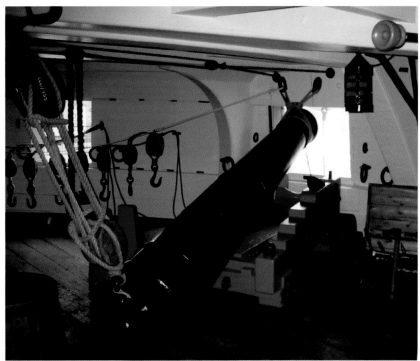

3oz of litharge, to be put in after the lamp-black and turpentine have been well mixed. Add 1oz of umber to give it a gloss, and 1 gallon of bright varnish.

- Compound for blacking guns. To 1 gallon of vinegar add ¼lb of iron rust, and let it stand for one week. Then add a pound of dry lamp-black and ¾lb of copperas, and stir it up at intervals for a couple of days. Lay five to six coats on the gun with a sponge, allowing it to dry well between each application. Polish with linseed oil and a soft woollen rag. It will look like ebony.
- Paint guns black with a standard mix of black paint using linseed oil and lead as a base.
- Coat guns with tar and tallow.

If the Gunner is directed to strike any guns into the hold he is to:

- Pay them all over with a thick coat of warm tar and tallow mixed together.
- Wash the bore of the gun with fresh water and very carefully sponge and dry the inside.
- Insert a good full wad, dipped in the same mixture, about 1ft within the muzzle.
- Drive the tampion well into the muzzle and seal it with putty.
- Drive a cork tight into the touch hole and secure it there.

Note: for further details see Article XXVII, Chapter 11, 'For the Gunner', *The Admiralty Regulations 1801–1809*.

Maintaining the gun carriages

The Gunner is periodically to inspect all gun carriages for obvious defects and report any failings to the Captain and the Ordnance Board. The Gunner must also make arrangements for making good repairs or replacement as necessary. Likewise, all gun carriages are to be thoroughly surveyed by the Inspector of the Ordnance Board. This is normally done when guns and carriages are embarked when the ship is storing for sea service. The Gunner, with the assistance of the Armourer, is to undertake the following maintenance procedures. The Armourer is to perform any smith's work at the forge for the repair of any iron fittings.

- Carriages are to be painted once annually by men directed by the Ordnance Board or alternatively by the ship's crew.
- Check all wrought-iron work for corrosion or signs of pitting. This is especially relevant to those carriages located on exposed open decks. Repaint or replace same as required. The Armourer is to manufacture or rework new at the forge as directed.
- Trucks are to be inspected for shakes (splits in wood grain) and replaced accordingly. It is recommended that trucks are turned 90 degrees weekly if guns have not been moved.
- Grease all axletree bearings and trucks weekly to ensure guns move back and forth with ease. Use slush from the galley or tallow. This action can help speed up gun drill.
- Check lynch pins (four in number) retaining the trucks are present and can be freely withdrawn. This action is important as it may be necessary to replace damaged trucks in action quickly (rather like a pit-stop wheel change) in order to maintain speed and continuity of gun drill.
- Check condition of stool bed and quoin(s).
- Check all forelock pins and respective roves are present; if not replace same.
- Check all forelock pins and respective roves are tight; if not re-drive respective bolts and forelock pins.

ABOVE Constant exposure to the elements means particular attention to guns and gun carriages above deck is required.

(Jonathan Falconer)

- Check axletree angled lock plates are secure; if not re-drive respective bolts and forelock pins.
- Check cap square forelock pins are present and retained with their chains; if not replace same and secure.
- Check cap squares are secure and bed down correctly; if not re-drive respective forelock pins.
- Check all eyebolts (four to eight in number) are secure; if not re-drive same.
- Check ringbolts on cheeks (two in number) are secure; if not re-drive same.
- Check transverse tie bolts for cheeks and transom are secure; if not re-drive same.

Notes:
1. Roves are flat washers.
2. Forelock pins are tapered cotter pins.
3. Slush is residue fat removed from galley coppers after boiling meat. This substance can also be used to grease the side tackle blocks to ease and maintain the speed of gun drill.

Repairs and action damage control

Although damage by enemy shot to the ship's upper works, broadsides, masts, spars, rigging and sails was significant, damage sustained below the waterline was far more critical because the volume of seawater flooding into the hold very much affected the ship's stability, speed and manoeuvrability. The *Victory* was making 12in (30.5cm) in depth of water in her hold after the Battle of Trafalgar with the pumps constantly manned. During battle the Carpenter and his mates are to be stationed below the waterline in the orlop wing passages to undertake immediate damage control, stopping up shot holes and shattered planking. The original orders directing this requirement for carpenters were first authorised in the Tudor Navy of Henry VIII, in ships like the *Mary Rose* when naval artillery became the predominant weapon. Tools used by the Carpenter's crew for stopping up holes comprise: pin mauls, hammers, large wooden mallets, hand saws, adzes and measuring battens. The latter are similar to the gunter battens used today in

modern naval damage control. The materials used to effect repairs include: conical wooden plugs, soft-wood wedges, sheets of lead to cover shot holes, flat-headed nails to secure lead sheets, pads of oakham used with the lead sheeting, wooden pads to cover shot holes or retain plugs and timber shores to hold sprung planking or retain wooden or lead pads. Other items employed to stop up holes can be anything at hand to form a pad, such as bedding, hammocks and clothing. Shot holes in *Victory's* upper works can be covered in with squares of canvas tacked in place or lead and copper sheeting. In short, the methods of damage repair used in 1800 differed little to damage-control methods applied in ships during the Second World War.

Capstan maintenance and repairs

The *Victory* is fitted with two double capstans, the main or centre-line capstan and the jeer capstan. The former is used for hauling the great anchor cables with a 'messenger' (see pages 74–5). The latter was mainly used for hoisting in stores, guns, boats, raising topmasts, yards and hoisting the lower yards on their 'jeer' tackle. The jeer capstan could also be employed to haul in the anchor cables should the main capstan be out of service by means of a cable called a vyol and vyol block.

Both capstans are virtually identical in construction, sharing a common spindle passing through two decks, the upper and lower sections working in unison with its counterpart on the deck above or below. Although both main and jeer are similar, the current main capstan is a replica fitted during the *Victory's* restoration in 1922–28, the original having been removed far earlier. However, the jeer capstan, as recently discovered, is original and still able to be turned in its original bearings. More importantly, it is the only surviving example of a late 18th-century capstan.

The word capstan originates from the Latin word '*capistrum*' and Anglo-Saxon word '*capster*', which mean 'halter'.

To reiterate, each single capstan unit is virtually identical in construction and comprises the following main components:

Capstan components

Component	Material
Lower partners	Oak
Main bearing	Forged iron
Spigot	Cast iron
Spindle	Oak
Pawlring	Cast iron
Pawlhead	Elm
Pawls	Cast iron
Whelps	Elm or oak
Chocks	Oak
Trundlehead	Elm
Muntins	Oak
Upper bearing	Forged iron
Upper partners	Oak
Drumhead	Elm
Capstan bars	Ash
Waterway	Elm

Lower partners: oak planking 20in wide and 18in in depth laid between the middle tiers of the lower gun-deck carlings. The ends of the planks are let down 2in on the adjacent gun-deck beams, thus they stand proud of the flat of the lower deck planking.

Main bearing: flat, circular, cast-iron saucer located and bedded into the lower capstan partners to receive the spigot.

Spigot: cast-iron, inverted bush driven onto the lower end of the spindle, located with radial screw bolts.

Spindle: made from a single baulk of oak about 13ft long and 2ft 4in in diameter. The lower half on the lower gun deck is fashioned 10-sided in cross section; the upper half on the middle gun deck is 12-sided. The sides formed provide good faying faces to receive the inner side of the whelps and the semicircular segments forming the trundlehead and drumhead.

Pawlring: cast in iron as a single component, it comprises two flanged rings, the outermost

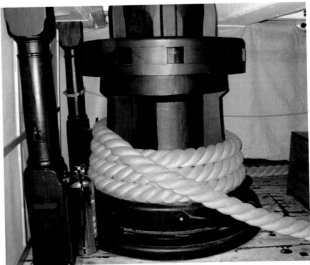

having an OD (outside diameter) of 5ft 7in and the innermost an OD of 4ft 7in; the width and depth of the rings is 6in. The annular space formed between the two rings is integrally furnished with iron stops 4in deep and 6in long, providing the ratchet mechanism for the pawls. The entire ring assembly is let down into a recess on the upper surface of the capstan

119

Pawls: these are four short, 4in cast-iron bars in square formation bolted to the pawlhead. Each is able to rotate about the axis of its bolts in order to reverse the ratchet effect.

Whelps: made of elm or oak, five are fitted to the lower capstan fayed and bolted radially to the spindle. Their function is to increase the 'sweep' diameter of the capstan; each is angled on their outer side to produce the surge for the capstan. Their upper ends are let into and fitted into the underside of the trundlehead.

Chocks: made of oak, these are fitted horizontally between the whelps. The lower set support the whelps against torque; the upper set serve the same purpose, but equally provide a land to bolt down the components forming the trundlehead.

partners and vertically bolted at every space between the stops.

Pawlhead: made from elm, this comprises a circular base for the lower half of the capstan on the lower gun deck. Formed in four segments, it is bolted radially to the spindle. It serves two functions: to support the whelps; and to house the iron pawls that are bolted around its periphery. Also inlaid into its periphery are iron plates with retractable bolts or pins, which when withdrawn protrude some 3in. These are used to keep the pawls disengaged from the pawlring when veering the capstan, letting the capstan run freely when not heaving a cable.

Trundlehead: made of elm, the whole assembly is formed from 4 semicircular pieces that fit over the 10-sided section of the spindle. The upper surface of the lower layer is radially cut with 12 tapered slots forming the sockets for the capstan bars. The underside surface is recessed to bed down upon and lock the whelps. The underside surface of the upper layer is radially cut with 12 capstan bar slots forming the other half of the capstan bar sockets. Once both layers forming the entire

trundlehead assembly are fitted over the spindle, they are jointly bolted with vertical forelock bolts passing through to the upper tier of chocks.

Muntins: made of oak, these comprise a series of vertical tapered battens nailed around the periphery of the spindle. Their lower ends terminate on the top face of the trundlehead, their upper ends level with the underside of the upper capstan partners. Their function is to lock the capstan vertically to prevent it from inadvertently lifting.

Upper bearing: this is a simple iron sleeve of the same diameter as the spindle, vertically inserted into the capstan partners of the middle gun deck. Although dimensionally smaller, these partners are fitted identically to those on the lower gun deck.

The upper half of the capstan has no solid base like the lower capstan, but does have elm whelps and oak chocks, which are fitted in the same manner as those previously described.

Drumhead: made of elm, this is manufactured in the same manner as the trundlehead, with the exception that the upper layer is fashioned with a 10-sided mortise to fit over the head of the spindle, producing a closed top.

Capstan bars: made of ash, these have one end tapered to fit into the sockets around the trundlehead and drumhead. Their outer ends are slotted to receive a rope called a swifter, which locks the bars in unison when used. The upper capstan has 14 capstan bars, the lower capstan has 12. With 10 men manning each capstan bar on both decks, a total of 260 seamen and marines could be employed to work the capstan.

Operation

The main (or centre-line) capstan: before operating totally, remove all of the wooden pillars surrounding the capstan. This can be done by flexing up the overhead beams using a beam jack. When the shallow tenon at the heads of the pillars have cleared the beam, knock down the pillars. You are also to take down and remove all ladders and the panelled

LEFT **Trundlehead, spindle and muntins of main centre-line capstan.**

ABOVE **Drumhead section of jeer capstan showing whelps, horizontal chocks and capstan bars fitted.**

LEFT **Detail of the jeer capstan trundlehead bar socket. Note the vertical bolt fastening the trundlehead to the horizontal chock between the whelps.**

partitions forming the wardroom bulkhead. Likewise, move any guns in close proximity in the way of the sweep of the capstan bars.

The jeer capstan: 1. Operating the upper half of the jeer capstan only, take down the iron pillars on the middle gun deck. This is done by driving their heels out of their iron shoes and knocking out the loose-fitted retaining bolts, or swing them up to their latches in the overhead beams. Ladders and guns in close proximity are also to be removed; 2. If jointly operating the upper and lower halves of this capstan, repeat as in 1 above but also remove the wooden pillars on the lower gun deck with a beam jack.

Maintenance

Although complex in their construction, capstans need very little maintenance. The most common problem is sprung (split) capstan bars, which are easily replaced. The only other maintenance involves greasing the bearings with 'slush' or tallow. Lubricating the upper bearing presents few problems as it can be easily accessed from above and below. Unfortunately, greasing the lower bearing is virtually impossible due to limited access under the pawlhead. Well beyond arm's length, this can only be successfully done by lifting the entire spindle with both capstans or by stripping down the upper capstan.

Getting out and refitting the capstan at sea

Prerequisites

- Ensure that the capstan not being worked upon is fully operable for anchoring and other tasks and prepare vyols, messengers, vyol blocks and necessary ropes to permit this.
- Remove hatch gratings from upper gun deck hatchways above capstan to be worked upon.
- Remove gratings laid upon the upper partners on the middle gun deck and any other obstructions.
- Prepare rope strops and lifting tackles for lifting the capstan.
- Prepare four long, stout oak timbers 9in sq in section for the lifting gear.
- Prepare four oak chocks 21in sq to support the capstan when lifted.

Procedure

- Lay two of the four timbers athwart the open upper gun deck hatchway coamings above the capstan head and rig the lifting gear.
- Knock out the forelock pins under the upper capstan head upper tier of chocks and remove roves to release the bolts.
- Knock all the forelock bolts upward to release the drumhead.

- Taking care to mark the mating surfaces of the 4 drumhead segments with the 12 faces of the spindle, carefully lift off the entire drumhead using the lifting gear if required. (Use a chisel and mark with Roman numerals.)
- Remove the muntins from the spindle above the trundlehead.
- Using sing mallets and soft-wood wedges, drive up and release the upper bearing and inspect same.
- Repeating steps 2, 3 and 4 above, remove the trundlehead carefully marking out the segments with the 10 faces of the spindle.
- Pass the prepared strops under the pawlhead and rig same to lifting gear.
- Using manpower hauling the lifting tackle falls, slowly hoist the lower capstan assembly and spindle as a single unit.
- When the unit is hoisted some 2ft 6in above the flat of the lower gun deck, tie off and secure the tackle falls and place the four oak chocks equidistant under the pawlhead and lay the two remaining 9in sq timbers across them. Once positioned, ease off the tackle falls and lower the capstan unit to rest with its pawlhead level upon the cross timbers and chocks on the flat of the lower gun deck.
- With the pawlhead resting 2ft 6in off the deck, sufficient room is now available to inspect, repair and/or repack the lower bearing with grease. This also provides an opportunity to remove any sand and filth deposited from the anchor cables that may bind the bearing.

Defective lower bearing: if it is found that the iron spigot or saucer forming the lower bearing is defective and needs replacing, only two avenues of repair are open unless the ship can be taken into a dockyard. These are: remove the entire capstan spindle; or lift the capstan assembly, whichever is convenient. The first is probably the most complex procedure as it involves removing more components, i.e., the pawlhead, trundlehead, drumhead and the upper and lower whelps.

To remove the entire capstan spindle:
- Remove the pawlhead by knocking out the

forelock pins on the lower tier of chocks of the lower capstan.

- Remove roves to release the bolts.
- Knock all the forelock bolts downward to release the pawlhead.
- Taking care to mark the mating surfaces of the 4 pawlhead segments with the 10 faces of the spindle, carefully lower and remove the pawlhead.
- Remove the trundlehead as described previously.
- Remove the drumhead as described previously.
- Using bolt extractors, withdraw the bolts fastening the upper and lower capstan whelps. Carefully split out the whelps from the spindle, ensuring that each are marked with corresponding numbers on the spindle.
- Remove the upper bearing as previously described.
- Take out and remove the upper capstan partners on the middle gun deck. (This involves considerable additional work.)
- Using lifting gear described previously, lift the entire spindle vertically up through the middle gun deck and lay aside to carry out repairs.
- Repair, replace or manufacture a new saucer and spigot on the forge as required.
- Replace saucer into its recess in the lower capstan partners.
- Refit the spigot onto the heel of the spindle.
- Pack the lower bearing with grease.
- Lower the spindle, ensuring that the spigot is correctly located in its saucer and that the spindle turns freely.
- Reassemble the upper capstan partners and insert the upper bearing and grease same.
- Check the vertical alignment of the spindle, ensuring that it turns freely.
- Commence reassembling the upper and lower capstans in reverse order as described above.

As it does not involve removing the whelps, lifting the capstan assembly is by far the easier task, providing that the heel of the spindle can be lifted sufficiently to give good access to carry out reasonable repairs. The only items to be removed are the pawlhead, trundlehead, the muntins and the upper bearing. The only

disadvantage is that the overall weight of the capstan is much heavier when lifting.

Pump maintenance and repairs

The *Victory* is fitted with two types of pumps: chain (or yard) pumps and elm-tree pumps. The former are used for pumping out the bilges in the hold to remove water from inherent leaks, etc. and the latter provide seawater for washing down decks, fire-fighting and domestic use. Like all items mechanical, they are prone to breakdown and require maintenance.

Elm-tree pumps

There are two fitted, one larboard and one starboard adjacent to the main mast. These are simple, hand-operated lift pumps taking a direct suction from sea for pumping up water for cleaning, washing down decks and fire-fighting. The pump casing, which is made from a single bored-out elm tree, passes directly out through the bottom of the ship: the upper end of the casing terminates on the lower or upper gun deck. The lower or suction end of the casing contains a 'fixed' valve box fitted with a poppet-type non-return valve. This valve box can be withdrawn for repair. A second valve box, similarly fitted with a poppet non-return, is free to reciprocate vertically within the elm-pump casing. This valve box is connected to a wooden brake handle via an iron rod known as the spear. The brake handle, made from ash, is supported by an iron crutch and located by a pivot pin.

Operation: because the elm casing remains filled with water up to level of the sea at the ship's waterline, the pump should not lose suction. The cyclic pump action is as follows:

- First movement. When the brake handle is pulled down the free valve box is raised. As it does so, its poppet valve will automatically shut permitting the water in the casing above the valve box to be lifted up to the discharge point.
- As the free valve box rises, a vacuum is created between the free and fixed valve boxes permitting the poppet valve in the

RIGHT Starboard
elm-tree pump on the
lower gun deck.
(Jonathan Falconer)

- Broken brake handle.
- Faulty or worn leather poppet valves.
- Seaweed fouling non-return poppet valves.
- Blocked suction inlet.

While the first three items are relatively easy to repair or replace, the final three require more attention.

Procedures:
For faulty or worn leather poppet valves:
- Disconnect spear from the brake handle.
- Using the spear, withdraw the free valve box.
- Inspect or replace the poppet valve.
- Using the spear, withdraw the fixed valve box.
- Inspect or replace the poppet valve.
- Replace valve boxes in reverse order using the spear.
- Connect spear to brake.

For seaweed fouling non-return poppet valves:
Carry out the same procedure described above for faulty or worn leather poppet valves and remove any material fouling the poppet valves and check that valves are undamaged and replace if necessary.

For blocked suction inlet, which could create quite a problem at sea:
- Remove the free and fixed valve boxes with the same procedure described above for faulty or worn leather poppet valves.
- Using a very long shaft of wood or iron (about 34ft larboard and 40ft starboard) passed down the pump casing, attempt to clear blockage. This can be more problematic with the starboard pump, which terminates on the lower gun deck, as there is little headroom clearance to insert a long shaft.
- If successful, withdraw the long shaft, replace the two valve boxes, spear and brake handle in accordance with the same procedure above for faulty or worn leather poppet valves.
- If unsuccessful, withdraw the long shaft, temporarily replace all fittings and put the

fixed valve box to open to allow in seawater to fill the pump casing above it.
- Second movement. When the brake handle is pulled up the free valve box descends compressing the water in the refilled part of the casing.
- On this action the poppet valve in the fixed valve box will automatically shut as a result of the pressure created above it, preventing more seawater entering the pump at the suction end.
- As the water below the free valve box is compressed, the poppet valve in the free valve box will automatically lift allowing water to pass through above the free valve box and the cycle is continued.

By continuously reciprocating the brake handle this pump can deliver 25 gallons (113.5 litres) of seawater per minute.

Maintenance: having very few moving parts, maintaining these pumps is relatively easy. Potential failings are:
- Disconnected spear.
- Broken pivot pin.

pump out of action and effect repairs when the ship is next docked down.

Main chain (or yard) pumps

There are four of these fitted, two in tandem to larboard and two in tandem to starboard adjacent to the main mast on the lower gun deck. Each individual pump comprises an elm cistern and all four serve two functions. First, to provide an overflow chamber temporarily to contain water discharged from the working chamber (described below). Secondly, to provide bearings to support the horizontal iron sprocket wheel-axle drive shafts set in the fore and aft plane to which horizontal crank handles can be engaged to rotate the pump. Fitted axially on the centre of each drive shaft is a circular sprocket wheel (or windlass) consisting of two flat circular plates 12in in diameter with a 6in hub. These are interconnected 12in parallel to each other with iron axial tie bars of 1in diameter. These bars serve as sprockets to engage the chain mechanism. Each of the cisterns is interconnected with square-section elm ducts or dales (wooden pipes) to convey water between them. The after two cisterns are fitted with square-section elm dales which convey bilge water to a large discharge scupper at the ship's side. To prevent ingress of rubbish, each cistern is covered by a detachable semicircular wooden hood of light construction. These are designed with sufficient clearance to ensure that the pump chain passes over the sprocket windlass without hindrance.

Descending from each cistern into the pump well in the hold are two metal and wooden pump casings. That fitted closer to the centre line is the back case and that fitted outboard is the return case (or working chamber). Each case terminates in the limber passage adjacent to the keelson in the hold.

The back case comprises a fabricated elm casing 10½in sq made in 3ft 6in sections connected by iron flanges and transition pieces fastened with forelocks (cotter pins). This facility provides access to repair the pump chain or replace sections of casing. When the pump is operating, the chain passes back down the back case to the limber passage.

The return case is constructed in a similar

way to the back case, with the exception that the lower sections are made of iron piping 7in in internal diameter bolted together with flanges. This portion, which actually serves as the working chamber of the pump, is 8ft 6in long. At the intersection point at the bottom end of the back case and the working chamber is an iron roller 6in in diameter which reverses the direction of the chain back up into the working chamber.

The chain, made of cast bronze, is the fundamental part of the pump's operation. It comprises a series of double and single flat links 7½in long joined together with ½in diameter bronze pins. Each link is fashioned with a 'hook' that engages with the tie rods of the sprocket wheel (or windlass). Every link is set at 36in distance (hence the term 'yard pump') and furnished with flanges, two

TOP Wooden upper sprocket wheel cover hood resting on pump dale interconnecting the pump cisterns. *(Jonathan Falconer)*

ABOVE Chain pump cistern discharge and starboard elm-tree pump. *(Jonathan Falconer)*

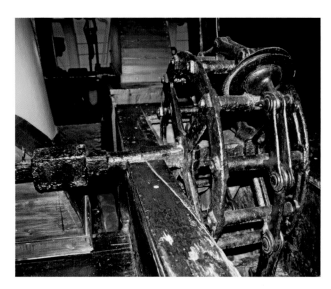

2. Lift and remove gratings from main hatchway and the hatch abaft the pumps.
3. Lift and remove tapered chock from the coamings of the aforementioned hatchways.
4. Where chock is removed, drop in, engage and slide in short deck boards athwart the ledges of the said hatches to fill up the hatch openings. The short boards form a temporary deck some 18in lower to make crank operation ergonomically easier for the men turning the cranks if they are standing on the flat of the lower deck.
5. Release hinged iron support brackets fixed from the overhead beams and lock in stays with their pins.
6. Fit up crank handles and extension pieces and lock all together with their forelocks.
7. Fit up discharge-pump dale to discharge scupper at the ship's side. This could be in the form of a large-diameter leather hose if the fixed wooden dale is not fitted.

When operating, the crank handles are to be turned towards the centre line of the ship, driving the chain to descend down the back case reverse direction and rise up the working chamber and return case lifting the water up to empty into the cisterns. If operating efficiently with four men manning the cranks, 1 ton (225 gallons) of water can be pumped in 44 seconds and with two men, 1 ton in 55 seconds. The chain pumps are to be operated daily after sounding the depth of water level in the pump well in the hold.

Maintenance: having many moving parts, maintaining this type of pump can be very labour intensive. However, despite this fact potential failings are somewhat limited. These are:

- Broken upper sprocket wheel.
- Broken lower sprocket wheel.
- Blocked suction.
- Chain disengaged from upper or lower sprocket wheels.
- Broken chain.
- Faulty or worn leather washers.
- Fractured saucers.

The first four are relatively simple to repair.

ABOVE Detail of the upper sprocket wheel and longitudinal connecting spindle.
(Jonathan Falconer)

1in-thick bronze circular saucers of 4½in diameter between which is sandwiched a 7in-diameter leather washer ¼in thick. The combined components are commonly called pistons. The clearance between the internal diameter of the working chamber and leather washer is marginal if not an interference fit. This is necessary because as the chain rises up through the return or working case these leather washers lift the bilge water to the cisterns. Each pair of pumps (larboard and starboard) is operated by four long iron cranks which engage into sprocket wheel-axle drive shafts and into each other. If using all cranks, they can extend 21ft 9in afore and 16ft 6in abaft the cisterns. Where necessary the cranks have extension shafts. When shipped the cranks and extension shafts are supported with fixed iron bearings mounted on deck pillars and by means of hinged iron support brackets fixed to the overhead beams. The linkage of all cranks and shafts are retained with iron forelocks (cotter pins).

Operation: the following are prerequisites:
1. Take down all ladders fitted in the fore and aft attitude in the way of the crank handles as they impair rotation of the cranks. Ladders that are fitted athwartships do not need removing as the cranks can rotate under or over the ladder styles.

Procedures:

1. For broken upper sprocket wheel:
   This is accessible and easy to replace, manufacture or repair on the forge. Before removing the sprocket wheel disengage and lift the chain and tie it just above the sprocket wheel for accessibility.

2. Broken lower sprocket wheel:
   This can be removed and repaired in the same manner as above, but is far less accessible at the bottom of the pump well.
   - If the well is quite full of water, operate the pumps on the opposite side of the ship to reduce the bilge water levels before undertaking the work.
   - The Carpenter is to enter and descend into the well via the grating or scuttle located on the starboard side of the mast room on the orlop, and remove the faulty lower sprocket wheel and replace the same on completion of repair.

3. Blocked suction:
   - Pump out the well using other pumps.
   - Carpenter's mate to descend into the pump well with bucket(s) tied to a line, lanthorn and suitable clearing tools and clear away all debris and check that lower sprocket wheel is not obstructed from turning.

4. Chain disengaged from upper or lower sprocket wheels:
   - Upper sprocket wheel: lift the chain on a length of small rope and realign the chain hooks back onto the upper sprocket wheel tie bars.
   - Lower sprocket wheel: Carpenter to descend into pump well and check and align the chain onto the lower sprocket wheel.
   - Fit one crank and turn the pump over two full cycles to ensure the chain is fully engaged with upper and lower sprocket wheels.

   Faults 5–7 require more attention.

5. Broken chain. If the chain has broken, it is highly probable that it has fallen down the return or back case and, consequently, the chain first needs to be retrieved.
   - Carpenter and mate to descend into the pump well with a bucket, lanthorn and

ABOVE Pump well showing mast step, working cases of larboard chain pumps, back case exposing chain and saucers, eight-sided casing of larboard elm-tree pump. *(Peter Goodwin)*

LEFT Cylindrical working chamber of the larboard fore chain pump. *(Peter Goodwin)*

suitable tools.
- Using a hammer, drive out the forelock pins locating the side panels of the square sections of the pump return or back case and remove same and iron retaining hoops to access the chain.
- Locate the sections of chain and fully withdraw them on a long hook or small rope.

- Lay out the chain along the lower gun deck or orlop and inspect all links and link pins for brakeage.
- Replace links or pins with spares as necessary.

Because the chain links and pins are made of cast bronze they cannot be effectively repaired at sea. While the chain is out of use, the opportunity is taken to inspect and replace worn leather washers, saucers, etc.

- Replace chain and align hooks onto upper and lower sprocket wheels.
- Replace side panels of the sections of the pump return or back cases and secure same with forelock pins and hoops as necessary.
- Fit one crank and turn the pump over two full cycles to ensure the chain is fully engaged with the upper and lower sprocket wheels.

6. Faulty or worn leather washers. Follow the same procedure given in point 5 above and replace components accordingly, making new as required from hides.

7. Fractured saucers. Follow the same procedure given in point 5 above and replace components accordingly from spares.

If none of items 5, 6, and 7 can be rectified, the pump is placed out of action. This is achieved through removing the upper sprocket wheel from its drive spindle by knocking out the forelock retaining pins and shunting the spindle afore or abaft, and then reconnecting the same to the corresponding drive shaft on the adjacent pump to enable drive cranks to be extended afore or abaft as required. It will also be necessary temporarily to stop up the openings at the top of the return and back cases or interconnecting ducts to prevent water from other pump cisterns flowing back down into the bilges.

Using the portable forge

The *Victory*, like many warships, carried a portable forge. This was used mainly by the ship's Armourer, who, equally skilled as a blacksmith, could undertake repairs and the manufacture of iron work, e.g., gun-port lid hinges, deadeye chains, boom irons and eyebolts.

The forge itself comprises a square iron frame of working height which supports the firehearth, under which is the hand-operated bellows which forces air up into the hearth.

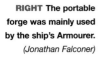

RIGHT The portable forge was mainly used by the ship's Armourer. *(Jonathan Falconer)*

The large bellows consists of two oak boards hinged at one end with thick leather sandwich between them, fastened with flat-headed nails. The bellows is operated by a lever mechanism connected to a wooden hand lever located on the left-hand side. The firehearth is covered by an iron hood closed off with two hinged doors. The top of the hood is furnished with a 4in-diameter iron stub pipe to which a copper flue and its elbows are connected. Operating the forge is no more difficult than using a forge on land with the exception that it presents a very serious fire hazard within the ship. To this end, many fire precautions are to be set in train before it is used. It should be set up in the waist of the upper gun deck, preferably under the larboard fore and aft gangboard, running between the quarterdeck and forecastle.

Before firing up the forge furnace the following prerequisites must be undertaken:

- Connect leather hoses to the discharge of the larboard elm-tree pump and lay out same to the forge and working area.
- Liberally wet down the flat of the deck surrounding the forge and the anvil on which the Armourer will be working.
- Scatter sand on the wetted deck areas.
- Place three buckets of sand near the forge and anvil.
- Place three buckets of water near the forge and anvil. Besides acting as a first-aid, fire-fighting appliance, the Armourer can equally use these for quenching and tempering the metals being worked.
- The pump and hoses are to be constantly manned by three men throughout the time the forge is in operation. The three 'firemen' are to continue manning the pump and hoses until the work is complete, the furnace has cooled and the coals and embers have been extinguished and removed.
- Throughout operation the fire is to be kept sufficiently low.
- If the weather and sea state suddenly worsen, the furnace fire is to be thoroughly extinguished.
- When work is complete the furnace fire is to be thoroughly extinguished.
- The entire working area, including the

LEFT The large bellows were hand operated by a lever mechanism.

(Jonathan Falconer)

deck area below, is to be searched for any potential fire hazard, which if found is to be eliminated appropriately.
- All coals and embers are to be removed into an iron bucket and very carefully lowered on a line overboard on the leeward side of the ship.

Red-hot shot

It is popularly believed that it was a common practice to fire red-hot round shot at enemy ships or fortifications. Unfortunately, this is a misconception for it presented too great a fire risk to be practical in a fleet battle. Consequently, this practice was not encouraged, primarily because it was often more dangerous to the parent ship than that of the enemy. Carrying and loading heated shot is itself a very dangerous evolution on land, let alone on board a wooden ship with tarred rigging and canvas sails.

Furthermore, it is doubtful if a forge of this limitation could produce the necessary temperatures to heat shot red-hot, and even if it could, the time taken would be impractical considering the high risks involved.

Chapter Six

Sailing a
Man-of-war

Born into an age before the
internal combustion engine,
the *Victory* was totally reliant
on manpower and wind
power for her operation.
To sail a large three-deck
ship-rigged vessel like the
Victory was a complex
activity that required
considerable skill and
teamwork from a crew of
some 800 men.

OPPOSITE The 28-gun frigate HMS *Surprise* enters a fog
bank. 'Standby to wear ship – up mainsail and mizzen spanker.'
(iStock)

Operating and manoeuvring a 3-decked sailing ship like the *Victory* differs little from any other ship-rigged vessel, i.e., having square sails set on 3 masts and a bowsprit, be this either a 2-decked 74-gun ship or a single-decked 28-gun frigate. The differences only relate to the physical size of the rig, masts and yards, and sail area, each carrying the same number of sails. The fundamental principles remain identical for each ship-rigged vessel, which still requires some 728 blocks aloft and the same number and type of ropes to operate the rig. Furthermore, no matter which ship you are in, the physical locations from where operating ropes are handled and secured are virtually universal for both physical and practical reasons. A seaman taken out of the *Victory* will equally know where the fore topgallant leechline fall is belayed in the frigates HMS *Trincomalee* (1817) and *Unicorn* (1824), likewise the American heavy frigate USS *Constitution*. This also applies to the iron warship HMS *Warrior* (1860).

Ship handling

Compared to modern yachts, sailing a large three-decked, ship-rigged ship is very complex, requiring considerable skill and many people working in teams responsible to each single mast and bowsprit. To achieve this will require:

- A highly qualified and confident Master (a singular role not to be confused with the ship's Captain).
- A lot of training at sail drill in order to get the teams working in unison in order to execute sailing manoeuvres swiftly.
- Very capable Topmen to hand sails aloft in all weather conditions, night and day.

As there are many permutations of how to handle a square-rigged ship under a host of variable conditions, for simplicity only the following major evolutions are described:

- Setting or taking in sails.
- Tacking ship.
- Wearing ship.
- Heaving to.
- Reefing sails.
- Getting the ship under sail from being at anchor.
- Bringing the ship to anchor.
- Sailing performance.

Setting or taking in sails

Before providing specific details of the various sailing evolutions listed above, this section gives simple explanations of the fundamental actions and ropes needed to set the square, triangular and quadrilateral sails regardless of the circumstances and requirements of the planned manoeuvre.

BELOW The author in earlier years when sailing in the Endeavour. *(Peter Goodwin)*

Square sails

Regardless of mast or yards these are all basically controlled or operated by the following ropes:

Control ropes for square sails

Item	Rope name	No.	Function
1.	Sheets	2	Hold down and out the outer lower corners
2.	Tacks	2	Hold down outer lower corners (course sails only)
3.	Clewlines	2	Haul up the outer lower corners into its yard
4.	Buntlines	3–4	Haul up the foot or bunt of the sail
5.	Leechlines	2	Haul up the leech or bunt of the sail to its yard
6.	Bowlines	2	Hold out and haul forward the leeches of the sail

In all cases related to the square sails and
the mizzen sail, all gaskets are to be untied
and let free and made up and tied clear once
the sail is set.

To set the main sail (or main course):
1. Let go and ease the clewlines, buntlines
   and leechlines.
2. Haul tight and set the sheets and tacks.
3. Make fast the clewlines, buntlines and
   leechlines.
4. Set the bowlines as appropriate to the
   wind conditions or tack sailing upon.

To set the fore sail (or fore course):
Repeat steps 1–4 above.

To set the topsails, topgallants, spritsails and royals:
Repeat steps 1–4 above with the exception
that none of these sails are rigged with
tacks and that spritsails and royals are not
rigged with bowlines.

Fore and aft triangular staysails and jib sails

Regardless of masts these are all basically controlled or operated by the following ropes:

Control ropes for staysails and jib sails

Item	Rope name	No.	Function
1.	Sheets	2	Hold down and out the after lower corners
2.	Tacks	1	Hold down the fore lower corners
3.	Halliards	1	Hoist up the sail on its stay
4.	Downhaulers	1	Haul down the sail
5.	Outhauler	1	Haul out the sail (jib sails only)
6.	Inhauler	1	Haul in the sail (jib sails only)

To set the fore topmast, main and mizzen staysails:

1. Let go and ease the downhauler, sheets and tacks.
2. Hoist the sail on its halliard and make fast the halliard.
3. Haul tight and set the sheets and tacks as appropriate to the wind conditions or tack sailing upon.

To set the fore jib and flying jibs:

1. Let go and ease the inhauler, sheets and tacks.
2. Hoist the sail on its halliard and make fast the halliard.
3. Haul tight the outhauler and heave out the traveller and make fast the outhauler.
4. Haul tight and set the sheets and tacks as appropriate to the wind conditions or tack sailing upon.

The fore and aft quadrilateral mizzen sail

This sail is controlled or operated by the following ropes:

Control ropes for the mizzen sail (spanker)

Item	Rope name	No.	Function
1.	Sheets	2	Hold down and haul out the after lower corner
2.	Tacks	1	Hold down the fore lower corner
3.	Brails	3	Hoist in and haul up the sail to its mast and gaff yard

To set the mizzen sail on its boom and gaff:

1. Let go and ease the brails.
2. Haul tight and heave out the sheet.
3. Haul tight and make fast the tack and make fast the sheet.

For taking in sails reverse the procedures given above.

Setting or taking in the steering (studding or stunsails):

This evolution is quite complex as it involves swinging out and setting the rigging of the lower stunsail booms, running out stunsails booms on the yards and hoisting the stunsail on their yards to the outer end of the booms.

For further information refer to: Darcy Lever, *The Young Officers Sheet Anchor* (1819); J. Harland, *Seamanship in the Age of the Fighting Sail* (1987); *Nare's Seamanship; The Kedge Anchor*.

ABOVE Armed
transport on a
starboard tack showing
main course and main
topsail set likewise
the mizzen staysail
(left), mizzen topmast
staysail stowed under
the main top and main
topgallant yard lowered
at the crosstrees.
(University of Chicago)

Tacking ship

A modern yacht, which by virtue of its fore
and aft rigged sails, is far abler to sail closer
to the wind (about 4 points, 46 degrees) than
a square-rigged ship like the *Victory*, which
is lucky if she can sail 6 points (69 degrees)
off the wind. Consequently, the only way to
sail on a course against wind is to carry out a
series of zigzag courses (called tacks) with the
wind prevailing first on one side of the ship,
i.e., larboard, and then upon the other, i.e.,
starboard, with the ship having to turn her head
through the wind at the end of each zigzag leg
in succession.

Advantages:
1. Tacking is a quicker and faster method
   of turning the ship through the wind onto
   another tack or course.

Disadvantages:
1. Tacking can put a lot of strain on the rigging.
2. The manoeuvre can often fail in a square-
   rigged ship if forward momentum of the ship
   is lost at the crucial moment.

*Before attempting to tack a square-rigged ship
the vessel must have sufficient way (momentum
forward) in order to turn the head through the
wind. Therefore, the manoeuvre must be precise
and quickly executed and failure to do so will
result in the ship's head falling back off the
wind and not getting onto the new tack. In this
situation, the ship is said to have 'missed stays'.
For the purpose of the written procedure the
ship is sailing close hauled (just off the wind) on
a larboard tack, i.e., the wind prevailing on the
larboard (port) side.*

Procedure

- Boatswain to call the hands standby to tack
  ship. All seamen go to their respective bracing
  station on deck and prepare and man the
  braces for running. Focslemen to stand by
  the jib and other fore staysails if set.
- Ease the helm to leeward (downwind) in order
  to increase the speed of the ship.
- When ready order 'Standby to tack'.
- Order 'Helm's a'lee'.
- Order 'Ready about'.
- Ease the jib and fore staysail sheets to assist
  the turn because as the wind spills from the
  square sails the speed of the ship is reduced.
- Order 'Haul taut main and mizzen sail haul'.
  Brace round the yards on the main and mizzen
  quickly to the opposite side in order to catch
  the wind. The sails are momentarily set aback
  driving the ship backward (sternboard).
- Put the helm amidships to centre the rudder;
  the ship loses way (slows down).
- By virtue of the wind direction the sails on the
  foremast are automatically set aback driving
  the ship's head through the wind towards the
  alternative tack.
- Ease the boom of the mizzen sail to larboard,
  and likewise the sheets of the jib and fore
  staysails ready to set on a starboard tack.
- Order 'Let go and haul'. Quickly brace round all
  yards on the foremast. The wind fills all sails on
  the fore, main and mizzen masts and the jibs, etc.
  bringing the ship's head onto a starboard tack.
- Focslemen man the weather (starboard) braces
  and brace the spritsail and sprit topsail yards to
  ease the strain on the bowsprit and jib booms.
- Trim all sails on other mast as required.
- Make fast and coil all sheets, tacks and braces
  and stand down the people.

LEFT *'It's a better master who knows when to take in more sail.'* With the wind hardening to a gale the escorting frigate (right) has already sent down her topgallant yards while topmen, laid out on the yards, take in a second reef to further shorten the fore and main topsails. Note the spritsail yard has been braced in to starboard to ease the strain on the bowsprit as the ship lays onto a starboard tack. *(From the painting 'Convoy off the Cornish coast' by Mark Myers)*

BELOW Tacking ship. *(Peter Goodwin)*

5 **'Focslemen** *man the weather (starboard) braces.'* Brace the spritsail and sprit topsail yards to ease the strain on the bowsprit and jib booms. Trim all sails on other mast as required. Make fast and coil all sheets, tacks and braces and stand down the people.

1 **'Stations for stays.'** Boatswain to call the hands standby to tack ship. All seamen go to their respective bracing station on deck and prepare and man the braces for running. *Focslemen* to stand by the jib and other fore staysails if set. Ease the helm to leeward (downwind) in order to increase the speed of the ship. When ready order *'Standby to tack'*.

WIND DIRECTION

2 **'Helm's a'lee.'** Order *'Ready about'*. Ease the jib and fore staysail sheets to assist the turn because as the wind spills from the square sails the speed of the ship is reduced.

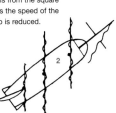

4 **'Let go and haul.'** Ease the boom of the mizzen sail to larboard, and likewise the sheets of the jib and fore staysails ready to set on a starboard tack. Order *'Let go and haul'*. Quickly brace round all yards on the foremast. The wind fills all sails on the fore, main and mizzen masts and the jibs, etc. bringing the ship's head onto a starboard tack.

3 **'Haul taut, main and mizzen sail haul.'** Brace round the yards on the main and mizzen quickly to the opposite side in order to catch the wind. The sails are momentarily set aback driving the ship backward (*sternboard*). Put the helm amidships to centre the rudder; the ship loses way (slows down). By virtue of the wind direction the sails on the foremast are automatically set aback driving the ship's head through the wind towards the alternative tack.

137

1 **'Stations for wearing ship.'** The crew to man their respective stations and prepare the braces for running. Once ready the next order, **'Stand by to wear ship'** is given.

2 **'Up mainsail and mizzen (spanker), brace in the after yards.'** The main course and mizzen sail are brailed up to prevent them opposing the turn. **'Up helm'**, the rudder turns the ship to starboard taking her stern across the wind.

3 **'Main and mizzen mast, let go and haul.'** The yards of the main and mizzen masts are braced round to feather the sails, allowing them to spill the wind.

4 As the wind comes onto the starboard quarter acting on the sails of the foremast and the head sail (jibs), the ship begins to turn.

5 **'Square the fore sails, let go and haul.'** The fore yards are squared and the headsails are hauled over to larboard.

WIND DIRECTION

8 The ship now proceeds on a starboard tack. All sails are trimmed and respective ropes coiled or secured.

7 With the sails on the main and mizzen masts filling, the ship gathers speed.

6 **'Haul aboard, haul out.'** The mainsail and mizzen sail (spanker) are re-set and the lee braces of the fore yards are braced round.

ABOVE Diagram showing the ship close-hauled on a larboard tack, with the wind coming from the larboard side.

(Peter Goodwin)

Wearing ship

The alternative manoeuvre to tacking is wearing, which involves turning the stern of the ship through the wind. In modern terms this manoeuvre is commonly called gybing.

Advantages:
1. The manoeuvre puts less strain upon the rigging.
2. Wearing is an easier and positive manoeuvre to undertake in a square-rigged ship to get onto the alternative tack or course.

Disadvantages:
1. A slower manoeuvre involving more sail handling.
For the purpose of the written procedure the ship is sailing close hauled (just off the wind) on a larboard tack, i.e., the wind prevailing on the larboard (port) side.

Procedure

- Boatswain to call the hands standby to wear ship. All seamen go to their respective bracing station on deck and prepare and man the braces, sheets, tacks and buntlines for running. Focslemen to stand by the jib and other fore staysails if set. Afterguard to stand by to handle the mizzen sail brails.
- Order 'Stand by to wear ship'.
- Order 'Up mainsail and mizzen'. Let go sheets and tacks of the mainsail and mizzen sail and let both free. Man the buntlines of the mainsail and brails of the mizzen sail and brail up both sails (sails are temporarily lifted and semi-furled) to prevent them opposing the turn.
- Order 'Up helm'. Put the helm to leeward taking the ship's stern across the wind.
- Order 'Main and mizzen mast let go and haul'. Brace round the yards of main and mizzen masts to feather the sails to allow the wind to spill from the sails.
- Order 'Square the fore sails let go and haul'.

Brace the fore yards square to the ship. Haul in the weather sheets of the jib and fore staysails, i.e., to larboard.

- Order 'Haul aboard, haul out'. Let go the buntlines and brails of the mainsail and brails of the mizzen sail and haul in on their sheets and tacks and reset the sails.
- Brace round and haul in the leeward braces of the fore yard ready for the next tack.
- With the sails on the main and mizzen mast filling the ship gathers speed on a starboard tack.
- Focslemen man the weather (starboard) braces and brace the spritsail and sprit topsail yards to ease the strain on the bowsprit and jib booms.
- Trim all sails on other mast as required.
- Make fast and coil all sheets, tacks and braces and stand down the people.

BELOW The frigate *Blenheim* paying off from her anchorage in Plymouth Sound. Downwind on a starboard tack, with the wind just on the starboard quarter, the ship is about to wear onto a larboard tack. The mizzen sail is brailed in, main and mizzen yards braced ready, and the main sail clewed up and hanging in its gear. Simultaneously the fore-yards are being braced round ready and the foresail is being set. Luff to windward the jib virtually does nothing until its sheets are passed over the stays when the stern goes through the wind..
(From the painting by Mark Myers)

Heaving to

This sailing manoeuvre is temporarily used to hold the ship still in the water for various reasons:

- Speaking to another ship.
- Lowering out or taking in the ship's boats when receiving boats from other ships in the squadron.
- Taking on a pilot.
- Taking a sounding with the lead.
- Retrieving a man who has gone overboard.
- Holding the ship alongside an enemy vessel when firing a broadside.
- Holding the ship to avoid going too close to the land or driving on a lee shore.

In principle, the ship will 'heave to' or be 'hove to' by counter bracing round the yards in such a manner that the wind plays against the sails on one mast setting them aback opposing the forward drive on the sails of the other mast. As a result, the ship is neither making way forward or going astern and virtually stops.

Procedure

- Focslemen to let go halliards and sheets of the jibs and fore topmast staysails, man the downhaulers and haul down the respective sails.
- Let go the fore course and main course sheets and tacks and furl both sails hauling in on the clewlines, buntlines and leechlines.
- Let go the fore, main and mizzen topgallant sheets.
- Man the fore, main and mizzen topgallant clewlines, and clew in the sails. This effectively reduces the downward thrust upon the ship's head.
- Put the helm a'lee.
- Man the main topsail braces and brace round the yard to the opposite setting of the yards on the other masts.
- With the main topsail now set aback (the wind playing on the fore face of the sail) the ship virtually stops in the water. In short, the wind effect on the ship is balanced.

BELOW Volunteers laid out along the yard untie the gaskets to let out the upper topsail of the square rigger *Elissa* in Galveston Bay, Texas. Built in Scotland in 1877, the *Elissa* served under both British and Norwegian mercantile flags until the early 20th century. Today she is owned by the Galveston Historical Foundation in the USA. *(iStock)*

Notes:

1. As an alternative to using the main topsail, the fore topsail can be put aback in the same manner as described in item 4 above. If, however, it is desired to back the mizzen topsail instead, this unfortunately involves bracing the crossjack, yard which could foul with the imposingly braced main yard.

2. Backing the fore topsail can be used to stop the ship suddenly to avoid collision.

3. A ship can also be said to be hove to when only having the main topsail set when the wind is blowing hard and steering by the wind, and likewise when keeping to the wind in gales with the storm staysails set.

Reefing sails

The purpose of reefing sails is to reduce the surface area of a sail in proportion to the increase in wind velocity as an alternative to taking in the whole sail. The facility to reef is provided by bands of plaited rope reef points set on canvas bands sewn on both faces of the sail parallel to the upper edge of the sail. The reef points are to be double the diameter of the yard to which the sail is bent. Reef cringles (rope eyes) are fastened at the end of each band at the luff (vertical edge) of the sail. These cringles provide a lifting point from which to hitch up the sail using the reef tackle suspended from the yard arm when taking in a reef. When reefing the sail is gathered up onto the top of the yard and secured with both sets of reef points, the two being temporarily tied over the top of the yard with a reef knot which can be quickly undone when taking out a reef.

The number of reef bands varies according to which type of sail. The courses (foresail and mainsail) have two reef bands, whereas the fore and main topsails, which have the greatest area, have four reef bands. The mizzen topsail being smaller has three and the spritsail has two set in the diagonal. The quadrilateral mizzen sail has three bands parallel to the boom and the lower and topsail stunsails have one. Although it is unlikely that both courses and topsails are reefed at the same time, for the purpose of the procedure given below the same rules apply, but whichever the case, the stunsail booms on lower and topsail yards do need to

be moved out of the way to give the Topmen room to work on the yards in safety and without hindrance to their task.

The following procedure applies to all three masts: the fore, main and mizzen, with the exception that there are no stunsail booms on the mizzen.

Procedure

- Order 'All hands reef topsails' as the Topmen go aloft into the tops and crosstrees.
- The men on deck let go the topgallant sheets and haul in the topgallant clewlines and clew up the topgallant sail to the mast head; or to ease the topgallant yard tye halliards and topgallant yard lifts and lower the topgallant yard.

LEFT The fore side of the *Endeavour*'s main topsail and roached bunt of the topgallant sail. Clearly seen are the three bands of reef points to shorten sail and respective reef cringles worked into the bolt rope at the leach of the sail. Also shown are the robbands lacing the sail to its yard; hanging in hanks from the yard are the gaskets for stowing the sail. Tying the two ends of the gasket over the yard and sail with a reef knot is a precarious task for the Topman, which involves leaning backwards to pass one end of the gasket back over the yard with one hand.

(Peter Goodwin)

LEFT The cap of the *Endeavour*'s main masthead. Note the ash ribs and lignum vitae trucks forming the yard parrel and the strop passing over the cap for the main yard lift blocks.

(Peter Goodwin)

- Men on deck next to ease the topsail yard tye halliards and topsail yard lifts and lower the topsail yard and at same time to haul in on the weather (windward) braces to spill the wind from the sail. Spilling the wind can be assisted by letting fly the topsail buntlines.
- Men on deck to haul in on the topsail reef tackles and man the stunsail boom tricing lines at the bitts.
- Order 'Trice up and lay out'. Tricing lines are hauled up raising the inner end of the stunsail booms canting them up at an angle giving room for the Topmen to lay out.
- The First Captain of the fore top to go to the weather yard arm and sit astride the yard, the Second Captain to go to the leeward yard arm.
- First Captain to haul out the weather earring of the sail and pass it to the man near him to reeve the earring through the reef cringle from aft to forward and pass it to the First Captain.
- When the First Captain orders 'Haul out to windward' the rest of the Topmen manning the yard are to reach down and take hold of the reef points to and all facing to leeward pull the sail to the weather side to allow the earring to be passed.
- The First Captain to take turns and secure the earring.
- The First Captain then orders 'Haul out to leeward'. All Topmen manning the yard are to reach down and take hold of the reef points to and all facing to windward pull the sail to the lee side to allow the lee earring to be passed and secured in the same manner by the Second Captain. When securing the earrings it is important that the sail is well aligned with the middle of the yard.
- All Topmen manning the yard are to reach down and haul the sail onto the yard and make fast the reef points using reef knots.

Notes:

1. When gathering in the sail the last fold is to be larger to cover the rest of the folded sail to prevent water getting in.
2. The second man at each end of the yard is to fold in the 'dog's ear', the section of canvas attached to the bight formed in the leech of the sail and lay it flat along the yard over the sail

and tuck it in before securing.

3. Care is to be taken to ensure that the lower reef points do not jam the run of the topgallant sheets.
4. Topmen to clear yard and return to deck. When yard is clear ease the stunsail boom tricing lines and reset the booms.
5. Men on deck to man the topsail tye halliards and lifts and hoist the topsail yard to its new position yard and haul taut the topsail sheets and reset the sail.
6. Men on deck to let go the topgallant sail clewlines, man the topgallant yard tye halliards and lifts and hoist the yard, haul in on the topgallant sail sheets and reset the sail.
7. Man the topsail and topgallant braces and brace round the topsail and topgallant yards to the position required according to the tack set.

Getting the ship under sail from being at anchor

Assuming that the *Victory* is fully armed, stored and provisioned and that she has a fully trained complement of 850 men and officers, the initial preparations for getting the ship under way are described below.

Prerequisites

1. All masts and yards are fully rigged with the standing and running rigging in accord with the rigging warrant.
2. All sails are bent to their respective yards, sheets, tacks, halliards, downhaulers, etc. and other associated rigging, i.e., clewlines, buntlines and bowlines.
3. All lower course yards, mizzen crossjack, spritsail and sprit topsail yards have been 'crossed' (hoisted).
4. All topsail and topgallant yards are ready to be crossed.
5. Ensign and jack staffs have been struck and stowed and respective flags flown at their halliards.
6. All prerequisites to weigh anchor have been carried out.
7. Steering gear is operable and rudder chock is fitted.
8. All signal halliards are rigged in order to

LEFT HMS *Surprise*, with the wind astern under plain sail and fore topmast staysail and jib set. Note that the replica clipper *Stat Amsterdam* following in her wake has split topsails, which were not used in men-of-war. *(iStock)*

communicate with other ships.

9. All lower and middle gun-deck gun-port lids are shut.
10. All guns are stowed and lashed for sea.
11. The ship's launch is afloat and manned to retrieve the anchor buoy and other arising requirements. All other boats are stowed on the skid beams or hanging in the davits.

Procedure

- Boatswains to call all hands to clear lower deck to prepare to get ship under way.
- Captains of the main, fore and mizzen tops go aloft to oversee rigging, manage Topmen and troubleshoot problems.
- Stand by to 'cross' the upper yards and man the main, fore and mizzen topsail yard tye halliards, parrel falls and lifts.
- Ease the topsail yard parrel falls, and let go the respective lifts.
- Hoist the respective topsail yards and make fast yard tye halliards and parrel falls. When hoisted they are said to be 'crossed'.
- Captain of the Tops to stand by to cross respective topgallant yards and by easing the topgallant parrel rope falls, let go the respective lifts.
- Hoist topgallant yards using their respective tye halliards and make fast the topgallant halliard and parrel falls.
- Once upper yards are crossed Topmen go aloft to lay out on the main and fore topsail yards and take off the gaskets ready to shake out the main and fore topsails. At the same time, Landsmen go aloft to lay out on the main and fore course yards and take off the gaskets ready to shake out the main and fore sails. There is no need to set topgallant sails at this stage.
- At same time focslemen to rig the starboard fish davit and lower the starboard cat block ready to cat, fish and stow the best bower anchor.
- 'Make sail.' Set main and fore courses, main and fore topsails. It may not be necessary to set the main and fore courses fully at this stage as they could impair visibility when manoeuvring in a confined anchorage.
- If required, shake out and set the main and fore sails.

- Man the halliards and set the main (or middle) and fore topmast staysails to steady the ship.
- Man the halliards and set the fore staysail or jib as required in order to provide steerage.
- When Topmen have cleared the yards and returned on deck man all course and topsail braces, and likewise stand by to hand respective sheets and course tacks and brace the yards round as required depending on which tack (larboard or starboard) the sails are to be set and the ship to proceed. Set the helm to bring the ship's head up onto the wind and make sail upon the anchor buoy. Depending on the confines of the anchorage, it may be necessary actually to drive the ship astern to break the anchor from the ground. Should this be the case then set the topsails aback and make sternboard until the anchor cable is either afore, i.e., the cable comes up at the same angle as the fore stay, or abaft, i.e., the cable comes up at the same angle as the main stay.
- Accepting that all prerequisites have been undertaken below decks, stand by to weigh anchor.
- Man the capstans with landsmen and marines and commence taking in the slack of anchor cable as the ship gathers way and works up upon the anchor buoy.
- Mizzen Topmen to go aloft, lay out and take off mizzen topsail gaskets and set the mizzen topsail.
- Shake out and sheet home the mizzen topsail and make fast the sheets on the mizzen topsail bitts.
- Afterguard men to stand by the mizzen topsail braces ready to set the mizzen topsail aback when the anchor is nearly 'up and down' (hanging vertically).
- Set the mizzen topsail aback to stop the ship. If the wind is blowing fresh and the ship is proceeding too fast, it may be necessary to set the main and fore topsails aback to gain some sternboard (move backwards astern).
- Alternatively, it may be necessary to go 'aback' in order to heave the ship to temporarily.
- Break the anchor from the ground and

bring it 'a hawse' (hanging just below the hawse hole).

- If the anchor does not break the ground, it may be necessary to trip the anchor using its buoy rope. This can be done using the windlass in the ship's launch, already on standby for such an evolution.
- Once the anchor is a 'hawse', brace the courses and topsail yards and sheet in the staysails and bring the ship onto the wind and bear away from the anchorage.
- Once under way, commence 'catting' and 'fishing' the anchor.
- Stow and lash the anchor in its chock.
- Unrig and stow the fish davit and its associated tackle, etc.
- Stow the anchor buoy in the starboard fore lower shrouds.
- After guard to check gaff parrel is free, ease the mizzen brails and set the mizzen sail.
- Stand by to man the mizzen boom sheets and set the mizzen on the appropriate tack.
- Hand the sheets and reset the fore topmast

staysails on the appropriate tack, likewise the jib sails if required.

- Focslemen to lay out on the bowsprit and lay out on the spritsail yard and take off spritsail gaskets ready to shake out the spritsail.
- Shake out and set the spritsail and sprit topsail if required.
- Man the weather spritsail and sprit topsail braces and set the respective yards 'a'cock' to ease the strain on the bowsprit.
- Man the foresail, mainsail and topsail braces and heave to.
- Bring the launch alongside to leeward and hoist in the boat.
- Make all plain sail and signal the ships sailing in company to come up on station two cables, length astern of the flagship.
- Stand by to tack and wear ship as directed.
- Focslemen to man the weather (starboard) braces and brace the spritsail and sprit topsail yards to ease the strain on the bowsprit and jib booms.

BELOW Armed Transport HMS *Bounty* sails broad in light airs under plain sail, with main and fore courses, topsails, and top gallants set, and likewise the main royal. *(University of Chicago)*

■ If the wind falls off, send the Topmen aloft to set the fore main and mizzen topgallant sails.

Note: do not set any additional sail unless wind dictates otherwise, but be prepared to take in or reduce sail quickly. It is a good Master who knows how much sail to set but a better one who knows how much to take in. As the darkness of night draws in it is best advised to take in topgallants and reduce sail or reef topsails. This avoids the danger of undertaking work aloft in darkness to reduce sail if the wind suddenly increases.

Bringing the ship to anchor

Unlike a modern vessel with mechanical power to drive the ship astern in order to pay out the anchor cable, this evolution will require the ship to be driven sternboard using wind power only and is consequently far more complex in its execution. While it is popularly conceived that the weight of the anchor holds the ship, it is the actual length and weight of the cable laid out along the seabed that is the major contributory factor in holding the ship, the anchor simply securing the cable to the ground. The amount of cable paid out for anchoring is generally to be three times the depth of water in which the ship is to be anchored. If the depth sounded is 10 fathoms (60ft), for example, then the length of cable required to hold the ship is 3 x 10 fathoms = 30 fathoms (180ft). This figure minus the depth of water of 10 fathoms = 20 fathoms (120ft), the length of which lies on the seabed from the anchor.

For the purpose of the following directions it is considered that the cable for the best bower is already bent (secured) to its starboard anchor.

Prerequisites

1. Clear lower deck of all hands to heave up the starboard anchor cable from the larboard cable tier on the orlop (cables are stowed on their opposing sides to allow easier cross-over amidships), and flake it out in long hanks (loops) up and down the length of the gun deck ready to run out through the starboard hawse hole.

2. If not already rigged, rig cable stoppers to starboard ringbolts near the two pairs of riding (mooring) bitts.

3. Take down and stow the starboard manger boards.

4. Reduce sail by letting go sheets and hauling in clewlines, buntlines and leechlines and take in the fore, main and mizzen topgallant sails and prepare the ship to heave to.

5. Station an able man in the starboard fore chains to take soundings with a lead and line to determine the depth of water at the intended anchorage.

Note: items 1–3 can be identically applied if using the larboard (or second bower).

Notes:

1. When taking soundings with the lead and line it is important to put tallow in the hollow at the bottom of the lead in order to pick up material content of the seabed, i.e., sand, shingle, shells, mud, etc. and to record the findings and depth in the ship's log. This is important as it provides information about the seabed to determine if the anchor will hold securely. The details should be transferred onto the chart for future reference.

2. Rig anchor buoy to the best bower anchor.

3. Launch the cutter from its davits.

4. Overhaul the falls of the starboard cat block ready to slip the anchor.

5. Focslemen man the sheets, clewlines, buntlines and leechlines and take in and furl the spritsail and sprit topsails if set. This will provide greater visibility in a crowded anchorage.

Procedure

■ Bring the ship's head round and turn to windward.

■ Let go the sheets and take in the main and mizzen staysails and brail up the mizzen sail.

■ Heave to by hauling in on the mizzen topsail braces and set the mizzen topsail aback.

■ Set the helm amidships.

■ Man the sheets, clewlines, buntlines and leechlines and take in and furl in the fore main and mizzen topgallant sails if set.

■ Let go foresail and mainsail sheets and tacks and haul in on clewlines, buntlines and leechlines and take in and furl both sails.

RIGHT Brig *Pilgrim* sails large in light airs on a starboard tack with courses, topsails, topgallants and royals set, also fore topmast staysail, jib and flying jib. *(iStock)*

set the sails aback. The ship will gather sternboard away from the anchor (drive the ship astern).

■ As the ship gathers sternboard, commence paying out the anchor cable estimating the length given out.

■ Ensure that the cable does not kink along the deck.

■ Once sufficient cable has been run out, heave to by bracing round the topsail yards to oppose each other to prevent the ship going farther astern.

■ Take compass bearings of fixed landmarks to determine the exact position of the ship, and record same, including the landmarks chosen, in the ship's log. This action is important as regular checks on the bearings taken will determine if the anchor remains holding or not.

■ Once anchored, man the sheets, tacks, clewlines, buntlines and leechlines and take in and furl all topsails on the fore, main and mizzen masts.

■ Topmen to go aloft and stow the sails and bend on their gaskets.

■ Focslemen lay out on the spritsail yards, stow the spritsail and sprit topsail and bend on their gaskets and stow or unbend the jibs and fore staysails.

■ Take further soundings afore and abaft and record same in the ship's log, likewise record the ship's draught marks. This is important for loading calculations and for trimming the ship if it is intended to embark stores, water and additional weights.

■ Set the anchor watch.

ABOVE The *Endeavour's* main top, topmast and crosstrees viewed from the deck. The main topsail and topgallant sails are set, and the main course has been clewed up to 'hang in its gear' so as to reduce driving the ship's head down in a rising swell of sea. Centre, the mizzen staysail is set to balance the ship to reduce roll and the mizzen topmast staysail has been lowered stowed under the main top to reduce sail. *(Peter Goodwin)*

■ Focslemen to let go the fore staysail and jib halliards and hauling in on the downhaulers, take in jibs and staysails.

■ Let go the anchor.

■ Man the fore and main topsail braces and

Embarking stores and equipment

Ballast

The manner in which the ship is to be ballasted and trimmed depends on the amount of provisions the ship is to be stored with for either four or six months. The actual amount of provisions to be carried is simply calculated by the purser by multiplying the total sum of individual allowances of food stuffs, beer and water, etc. per man by the number of men borne in the ship, then multiplied by the number of days in four or six months. As shown in the 'Observations of the Quality of His Majesty's Ship Victory', the ship is ballasted with 57 tons of iron and 200 tons of shingle (see p. 000).

The ingots of pig-iron ballast, each marked with the Government 'broad arrow', are to be laid out in the bottom of the hold. This ballast is supplied from the dockyards in three sizes and weights:

1. 3ft x 6in x 6in = 1½cwt (168lb)
2. 18in x 6in x 6in = 1cwt (112lb)
3. 12in x 4in x 4in = 1½cwt (56lb)

The 57 tons of iron listed above amounts to 760 ingots of pig iron weighing 1½cwt each. The 200 tons of shingle, which is to comprise pebbles of no greater than 1½in to 2in in size, served two functions:

1. To form a bed on which to set down the water casks to prevent them moving when the ship rolled.
2. Compensating ballast, which could be shifted in baskets to alter the trim of the ship as provisions, water, stores, shot and powder were consumed.

Embarking provisions and trimming the ship

The manner in which provisions are stowed can very much affect the trim and safety of the ship. The important responsibility of stowing the hold and trimming the ship is to be given to the ship's Master and the First Lieutenant. To prevent the casks shifting to one side of the ship, which could be disastrous at sea, the Carpenter and his crew are to fix the casks with a lattice of timber supports using dunnage (lengths of scrap timber).

When embarking provisions, water and beer from lighters sent from the victualling yard, the casks are lifted on board using the yard-arm tackles of the main and fore-course yards, and then using the stay tackles on the main stay are dropped through the main or fore hatchways. Smaller barrels can be parbuckled up the ship's side using the parallel fenders or passed through the lower gun-deck ports. The latter method is preferable when taking on powder. When provisioning the contents can rise in layers, known as 'tiers', to the full height of the hold, which is actually 15ft 6in (i.e., depth of the hold from the underside of the gun deck given as 21ft 6in minus 6ft for the height of the orlop).

Watering ship

The lower tiers in the hold contain the largest barrels know as 'leaguers', and each contains 150 gallons of water. When storing ship with 380 tons of water for 6 months a total of 568 leaguers is required. This amount of water in effect acts as ballast, and therefore as fresh water is consumed the casks are to be filled with salt water to compensate measurably for the weight loss.

As each full leaguer weighs approximately ⅔ of an Imperial ton, taking on water is extremely heavy work. To make loading easier the short-length portable planks forming the decking of the orlop above with their minor deck supports (ledges and carlings) are easily unshipped (removed), giving access down between the orlop beams. All the leaguers are laid in rows fore and aft, each with its bung up to minimise leakage. Most of the water is to be stowed in the after hold where there is less iron ballast.

Other wet provisions form the lower tiers in the main hold over which dry provisions, such as 45 tons of salted meat, pork, beef, suet, butter, raisins and peas, etc., are laid in tiers, while smaller casks are stowed in the side wings. Other foodstuffs kept in sacks, such as cabbages, pulses and limes, etc., are placed over the upper tier of barrels. All casks are to be marked with their contents and quantity, i.e., beef 70 pieces, raisins 300lb, etc.

Beer (50 tons), which compensates for water as it lasts longer, can be stowed in what used to be the fish room (fish no longer being carried). This amount of beer lasts 850 men 14 days. Spirits, brandy, rum and wine are to be stored and locked in the spirit room. Bread, in biscuit form (45 tons), and flour is to be stored in the tin-lined bread room.

Replenishment at sea

By 1800 replenishment at sea (today called a RAS) from victualling ships or escorts was very common. Consequently, this system, first started in the Seven Years War (1756–63), initiated what is today recognised as the Royal Fleet Auxiliary Service. Alternatively, boats were often sent inshore with casks to water ship, to purchase livestock, bullocks and sheep, to supplement the salted meat supplies, and to take on vegetables.

Embarking guns and gun carriages

By virtue of their weight embarking the guns is perhaps the most difficult and dangerous evolution, with the exception of getting masts in. Illustrations contemporary to the period indicate that guns could be brought over the ship's side using the yard-arm tackles rigged on the fore and main yards and then transferring the gun to the main-stay tackle rigged above the fore and main hatches. Although this method appears reasonably practical for the lighter 12-pounder guns, lowering and moving a 32-pounder gun of some 2¾ tons deadweight fore and aft between decks is highly precarious and presents considerable problems. Guns are generally conveyed out to the ship in a barge, which is tied up alongside the ship and with both vessels in motion from the sea state the easiest and safest method of loading guns is

to swing them into the ship, breech first, through the open gun ports directly onto their carriages. Moreover, it is best that guns are loaded into the ship from both sides simultaneously to maintain the balance and trim of the ship.

Prerequisites

1. All gun port lids to be fully opened and secured.
2. All gun carriages are to run up to the ships side and well secured at their ports with their tackle.
3. Place soft wood wedges under the trucks to prevent the carriages moving backward.
4. Remove gun carriage cap square cotter pins and hinge back cap squares to leave the trunnion recesses accessible.
5. Check that the stool bed and quoin are available to fit under the gun breech.
6. Prepare slings for the guns.
7. Check yard arm tackle safe to use.
8. Check stay tackle safe to use if required.
9. Check that the jeer capstan is ready for use.

10. Lower the main- (or fore-) course yard and run the tackle falls to the jeer capstan and check that the fall is well passed around the capstan whelps to prevent slip.

Procedure

- Pass the lifting sling around the gun barrel at the trunnions, which are its point of balance.
- Hook the yard-arm tackle to the lifting sling and mouse the hook.
- Secure two guide ropes to the breech ring of the gun to control and manoeuvre the gun during the lift.
- Using the jeer capstan, carefully lift the gun from the barge and sway the gun over.
- To counteract the weight the of the gun, rig the opposite yard-arm tackles with their fall blocks hooked into eyebolts fitted into the channels on the opposite side of the ship, or likewise use the yard (battle) preventer braces.
- Using the control ropes, manoeuvre the gun in through the gun-port breech end first.
- Pass one guide rope through the gun port into the ship where other men on the gun deck can help the gun onto its carriage.
- Slowly lower the gun, ensuring the trunnions settle down into their recesses and that the breech of the gun is fully lowered onto the hind axletree.
- When bedded down, close over the cap squares and lock same with their cotter pins.
- Remove soft-wood wedges from under trucks and remove the two guide ropes.
- Pass the gun breeching rope through the breech ring and hitch the ends to ringbolts at the ship's side.
- If loading a 24- or 32-pounder, rig the gun-preventer breeching ropes and hitch their ends to their designated ringbolts at the ship's side. Preventer breechings are not required for 12-pounder guns.
- Using the side tackles and muzzle lashings, stow the gun secure for sea passage.

If lowering the guns down through the main or fore hatchways from the main stay tackles, the same basic rules given above apply.

Hoisting in and getting out the ship's boats

Procedure

- If bringing the boat alongside the leeward side of the ship, lower the respective leeward main, and fore-yard tackles and hook them into the lifting ringbolts in the launch.
- Run the main-yard tackle falls to the jeer capstan.
- Man the jeer capstan and hoist in the launch by hauling in on the main-yard tackle falls.
- At the same time, manually hoist in the launch using the fore-yard tackle falls.
- To counteract the weight of the boat, rig the weather yard-arm tackles with their fall blocks hooked into the weather channels or likewise use the yard (battle) preventer braces.
- Man the main- and fore-yard braces and brace the yards to swing the launch inboard over the ship's waist. When swaying in the launch man the launch bow and stern painters to steady the boat.
- Transfer the weight of the launch onto the stay tackles rigged to the triatic stay set up between the masts just under the main and fore the tops or the main stay.
- At the same time as easing off the yard tackle falls and stay tackle falls, slowly lower the launch down onto the skid beams.
- Chock the launch.
- Unhook the stay and yard tackles from the launch and channels as appropriate, and using the tricing lines hitch up the main- and fore-yard tackles to their yard arms.

Getting the boats out and into the water from their stowage on the ship's skid beams fitted across the ship's waist is virtually a reverse process to that given above.

Hoisting in and lowering the boats slung in the quarter davits is a far simpler evolution, very much reliant on manpower hoisting up or lowering the boat on its falls rigged to the outer ends of the davits. The only other work involves raising or lowering the davits on their topping lifts running through blocks hooked to eyebolts fastened to the poop deck and strapped under the mizzen mast top.

Sailing performance

Well designed with good hydrodynamic lines below the waterline, the *Victory*'s sailing performance is quite exceptional for a first rate ship displacing 3,500 tons.

As detailed in the table below, the *Victory* could sail at a speed of 11 knots (12¼mph). This was one of the qualities that made the *Victory* very much sought after by various admirals. The records of the *Victory*'s sailing performance held in The National Archives read as follows:

Observations of the quality of His Majesty's Ship *Victory*, being given this 26th Day of November 1797

Her best Sailing Draft of Water when Victualled and Stored for Channel Service or as much lighter (at the same Difference) as she is able to bear Sail.	*Afore 23 Ft. 2 Ins.* *Abaft 24 Ft. 0 Ins.*
Her Lowest Gun deck Port will be above the surface of the Water	*5 Ft. 6 Ins.*
In a Topgallant Gale	*Behaves well, runs between 7 and 8 knots*
In a Topsail Gale	*Do as above will run 6, 7 or 8 knots*
Query the 1st How she Steers, and how she Wears and Stays	*She steers remarkably well, Wears very quick, and Seldom miss Stays when any other Ship will Stay.*
Under her Reeft Topsails *Circumstance.*	*Sails 4, 5 or 6 knots* *We have always found her to behave well in each*
Under her Courses	*Sails 2, 3 or 4 knots*
And Query, Whether she will Stay under her Courses	*Never Tryed her*

Query the 2nd.
In each Circumstance above mentioned (in Sailing with Other Ships) in what Proportion she gathers to Windward, and in what proportion she forereaches, and in general her Proportions in leeway.

In Sailing with Other ships, She holds her Wind very well with them, and forereaches upon most, or all Ships of three decks we have been in Company with, She makes Leeway from ½ a point in pretty Smooth water and increases as the sea rises, to 4 or 5 points under her Courses in a Great sea.

Query the 3rd.
How she proves in Sailing thro' all the variations of the Wind from its being two feet abaft the Beam, to its Veering forward upon the Bowline in every Strength of Gale, especially if a stiff Gale and a Head sea, and how many Knots she runs in each Circumstance and how she carries her Helm.

With the Wind Large she Sails well, she does with the Wind Veering forward to being close haul'd and forereaches upon most ships, but in a head sea she seldom more than holds her own with them, and will run from 10 or 11 knots Large to 4 to 3 knots Close haul'd on a head Sea. Carries her helm generally ½ a turn of the wheel or Something less aweather, but in a head sea at times a little alee.

Query the 4th.
The most knots she runs before the Wind;
and how she Rolls in a trough of Sea

10 to 11 knots – Rolls Easy and Strains nothing.

Query the 5th.
How she behaves in Lying Too, or a Try, under
a Mainsail, also under a Mizon balanc'd

Lays too well, Especially under storm staysails.

Query the 6th.
What a Roader she is, and how she careens

She Rides rather heavy on her anchor, never Careen'd her.

Query the 7th.
If upon Trial the best sailing Draft of water given
as above should not prove to be so what is
her best sailing Draft of Water

The above have found to be the best.

Query the 8th
What is her best Draft of water when victualled
for six Months, and Stored for Foreign Service

Afore 24 Ft. 3 Ins.   *Abaft 24 Ft. 6 Ins.*

Query the 9th
What height is her lowest Gun-deck Port above the Surface of the Water

4 Ft. 0 Ins

Query the 10th
Trim of the Ship

10 Inches by the Stern

Query the 11th
How she stands under sails

*She is thought to be a Stiff Ship; easily bought down – as
certain bearing and after that sufficiently Stiff*

Query the 12th
The Quantities of Iron and Shingle Ballast on Board

57 Tons of Iron, and 200 tons of Shingle

Query the 13th
How she stows her Provisions and water and What Quantity of the latter she carries with Four to Six Months
provisions; also the Quantity of Shingle or Iron Ballast which may be put out when she is victualled for Six Months.

*She will stow Four Months Provisions in the aft {hold} but when Six Months is ordered the Great part of the Wet
provisions are obliged to be put in the Main hold with four Months provisions; Will carry 380 tons of Water and Six
Months 335 tons. Bread room will just stow six months bread loose.*

Query the 14th
The Weight of the Provisions taken on Board,
in Consequence of being stored for the above time

Suppose 6 Months Provisions to be 300 tons.

Not having received a plan of the hold from the former Master – the stowage of Iron and shingle

Will. Cummings Capt

(Reference source: TNA, Adm 95/39/76)

Chapter Seven

Manning the *Victory*

The *Victory*'s crew comprised many ranks and rates, with about 90 per cent actually involved in working the ship, manning the guns and operating the rigging and sails. Most commissioned officers and senior warrant officers lived in cabins, but for the ship's company who inhabited the cramped lower decks it was a crowded ship.

OPPOSITE Marines' hammocks and mess tables on the middle gun deck, with sea chests underneath for seating. *(Jonathan Falconer)*

The official complement for a first rate 100-gun ship is 850, but like many ships at the time *Victory* would be marginally undermanned. Excluding the 15 commissioned naval and marine officers, the midshipmen, warrant officers, and those men making up the supply, secretariat and the Admiral's retinue (86 in total), the number of men working the ship and manning the guns and operating the rigging and sails is 719.

A high number of men is required because without powered machinery every task involved with operating the ship requires hard physical labour from large teams of people. Gun-crew numbers are calculated according to overall weight of a gun and its carriage, i.e., one man for every 500lb. Consequently, gun crews have to be divided between two guns when firing from both sides of the ship, otherwise the total complement to fulfil this requirement is unrealistic both from the perspective of living space and provisioning.

Victory's complement

Victory's complement of personnel comprised many ranks and rates as shown in the following table:

Rank and status	No.
Admiral Lord Nelson	1
Captain Thomas Hardy	1
Commissioned naval officers (lieutenants)	7
Commissioned Royal Marine officers	4
Senior warrant officers: Master, Gunner, Boatswain, Carpenter and Purser	5
Secondary warrant officers: Chaplain, Surgeon and ship's Cook	3
Midshipmen	21
Supply, secretariat and Admiral's retinue	42
Petty officers and miscellaneous mates	71
Able seamen	212
Ordinary seamen	193
Landsmen	87
Marines: sergeants and privates	142
Boys (aged between 12 and 19)	31
TOTAL	**820**

Excluding the Admiral and Captain, who have total privacy, most of the commissioned officers share individual cabins in the wardroom, sleeping in cots. With the exception of the Master (the most senior warrant officer), with his own cabin with its cot adjacent to the Captain's quarters under the poop, most of the senior warrant officers holding key posts within the ship's organisation generally live in cabins with fitted bunks on the orlop. This group comprises the Surgeon, Purser, Captain's Clerk, Boatswain and Carpenter. The Gunner is the exception as he usually berths in the gun room at the after end of the lower deck. The *Victory*'s standing officers, the Master, Boatswain, Carpenter and Gunner, generally remain in the ship when the ship is taken out of commission to act as ship keepers. If the ship is to sail as a 'private ship', i.e., not carrying an admiral, the Captain can transfer into the Admiral's quarters and give over his own cabin under the poop to the First Lieutenant.

The remainder of the ship's company live on the gun decks with the majority of the seamen on the lower gun deck, marines on the middle gun deck, midshipmen and some of the mates on the orlop and the boys mainly in the cable tiers. In the *Victory* the hammocks, suspended from battens nailed to the beams, were 16in apart. Although this appears quite crowded, it must be remembered that a good part half of the ship's company will be on watch at any one time adding a further 16in space per man. Although conditions in the Georgian Navy may seem harsh by today's standards, hammocks and broadside messing arrangements were still common in naval ships just 50 years ago.

Organisation

To run the ship efficiently you are to divide the ship's company and lieutenants, etc. into two watches, named appropriately to each side of the ship: Larboard and Starboard. The name larboard for the port side of the ship was not formally introduced by Admiralty Order until 1849.

Divisions

The Larboard and Starboard watches that worked alternative shifts were further separated into divisions representing different 'parts of ship', where each man has a dedicated station and will undertake different tasks depending on which sailing evolution was being ordered. These duties are set down in the ship's Watch and Station Bill, providing a formal ship's routine and allocation of manpower. This is to be drawn up by the ship's Captain, Master and First Lieutenant. In brief, the men are grouped accordingly to work each mast or particular deck areas of the ship. The divisions within the ship are to comprise the Foretopmen, Maintopmen and Mizzentopmen, Forecastlemen, Waisters and the Afterguard. For welfare reasons each division is to be overseen by an appointed officer and a midshipman.

Topmen

These are to be the youngest and fittest able seamen and will man the uppermost yards on each mast. Robert Wilson, a common seaman of the time, summarised the necessary qualities to be a Topman: '[He] not only requires alertness but courage, to ascend in a manner sky high when stormy winds do blow; in short they must exert themselves briskly. The youngest of the topmen generally go highest.' Captain Marryat wrote that Topmen are 'the smartest able seamen' and recommended that a 36-gun frigate should have 20 Foretopmen and 26 Maintopmen, of which 2 are to be petty officers or able seamen, the others to be training landsmen. In charge of each group is to be an experienced petty officer called the Captain of the Foretop, Maintop or Mizzentop accordingly. For the *Victory* it is suggested that you will need 24 Foretopmen and 32 Maintopmen.

Forecastle-men

These were overseen by the Captain of the Forecastle, a petty officer holding the post of Boatswain's Mate. These men, who are generally the best but oldest highly skilled seamen in the ship, mainly work with the gear associated with the anchors and operate the sails on the bowsprit.

Waisters

This division is to be stationed in the centre of the ship, i.e., the ship's waist, between the

forecastle and quarterdeck. They mainly comprise the landsmen, who, with lesser seamanship skills, add weight and power for hauling up or bracing round the yards of the fore and main mast. They are also to provide manpower for hoisting out the ship's boats, which are housed on skid beams crossing the ship's waist. In time, most landsmen can become competent seamen, the youngest working aloft, the others elsewhere as required.

ABOVE Main mast activity on the *Elissa*. *(iStock)*

Afterguard

This remaining division are to work on the quarterdeck and poop manning the braces, sheets, tacks, halliards, clewlines, buntlines and leechlines as required. They rarely go aloft unless to assist furling the main sail. These men also work setting the large fore and aft quadrilateral mizzen sail set on the mizzen mast. Men picked from this division can also assist with the signal halliards.

Daily routine and watches at sea

When at sea each day is to commence at 12 o'clock noon when astronomical sightings are taken to determine the ship's position. It is from this time that the daily routine is to be divided into a seven-watch system (not to be confused with the Larboard and Starboard divisions of seamen.) Each watch and its time span are to be as follows:

Notes:
1. The reason why two dog watches were created out of one 4-hour period is to ensure that the routine altered daily, otherwise the men of each watch would keep the same hours indefinitely.
2. The origin of the term dog watch is uncertain but it may correspond to the rising of the Dog Star Sirius, or it may relate to laying low or 'dogo' for half the standard watch hours to switch watches as stated in previous note.
3. In the Royal Navy the Second Dog watch was later renamed Last Dog, the title Second Dog being retained in the Mercantile Service.

Ship's time

This is to be notified by striking the ship's bell mounted in its belfry at the after end of the forecastle. Time is measured using an hour and a half-hour sand glass, the timekeeper (usually a marine) is to man the bell and ring it every half-hour, the number of rings denoting actual time. Taking the afternoon, dog and first watches, for example, the bell is rung as follows with eight bells terminating the previous watch:

The Ship's Watch system

Watch title	Time period (standard time)	24-hour time period	Origins of title
Afternoon	12 noon to 4:00pm	12:00 to 16:00	Self-explanatory
First Dog	4:00pm to 6:00pm	16:00 to 18:00	Associated with Dog Star Sirius
Second Dog	6:00pm to 8:00pm	18:00 to 20:00	Associated with Dog Star Sirius
First	8:00pm to 12 midnight	20:00 to 00:59	First watch of the night
Middle	12:00am to 4:00am	00:01 to 04:00	Middle of the night
Morning	4:00pm to 8:00pm	04:00 to 08:00	Self-explanatory
Forenoon	8:00am to 12 noon	08:00 to 12:00	Self-explanatory

Ship's time

Watch title	Standard time	24 hour time	Number of rings
Afternoon	12:00 noon	12:00	8 bells
	12:30pm	12:30	1 bells
	1:00pm	13:00	2 bells
	1:30pm	13:30	3 bells
	2:00pm	14:00	4 bells
	2:30pm	14:30	5 bells
	3:00pm	15:00	6 bells
	3:30pm	15:30	7 bells
First Dog	4:00pm	16:00	8 bells
	4:30pm	16:30	1 bell
	5:00pm	17:00	2 bells
	5:30pm	17:30	3 bells
Second (or Last) Dog	6:00pm	18:00	8 bells
	6:30pm	18:30	1 bell
	7:00pm	19:00	2 bells
	7:30pm	19:30	3 bells
First	8:00pm	20:00	8 bells
	8:30pm	20:30	1 bell
	9:00pm	21:00	2 bells
	9:30pm	21:30	3 bells
	10:00pm	22:00	4 bells
	10:30pm	22:30	5 bells
	11:00pm	23:00	6 bells
	11:30pm	23:30	7 bells
Middle	12:00am	00:01	8 bells
Etc	Etc.	Etc.	Etc.

Commanding a first rate ship of the line

Officers appointed to command a ship of the line were generally well experienced men who had proved themselves within the Service, or who were appointed as a protégé by an admiral. Nelson himself took Captain Thomas Hardy into the *Victory* with him when taking command of the Mediterranean fleet in July 1803. The *Victory*'s previous commander, Samuel Sutton, was transferred into Hardy's previous (lesser) command of the frigate *Amphion*. However, there were exceptions. John Bazely, who temporarily commanded the *Victory* in 1778, had previously gained recognition after capturing the American 18-gun commerce raider *Lexington* when commanding the 12-gun armed cutter HMS *Alert* the previous year.

Role and qualifications of a captain in the Georgian Navy

The Captain had overall supremacy in a man-of-war, overseeing all commissioned officers, warrant officers, petty officers, seamen and marines. On entering the ship he would formally read out his warrant to the entire ship's company, given to him by the Admiralty, authorising his command. Through his officers he would maintain discipline and good order throughout the ship. Each Sunday he would muster the entire ship's company to attend a Christian service after which he would read out the Articles of War to remind the men under his command of their duties and regulations.

It is likely he would have had experience in the Royal Navy from the age of about 13, working his way up from Midshipman to Lieutenant in various ships. During these formative years he would have learnt the basics of seamanship, sail handling, mathematics, trigonometry and a considerable amount of navigation.

Having taken his lieutenant's exams at about the age of 18, he would formally attain his status as a competent commissioned officer in the Royal Navy, taking charge of the ship on watch, taking charge of the gun deck gun batteries, and being appointed as a divisional officer responsible for the welfare of divisions of

men in the ship.

His next step on the promotion ladder was to be given command of a small ship before being promoted from Commander to Captain. After this he could be singled out for promotion to Post Captain. This rank indicated that an officer had attained considerable merit in command ability and initiative, and had been well tested in combat situations. Consequently, such men were generally notable enough to ascend to the flag rank of Commodore and Admiral.

The seaman's view

It is popularly believed that Georgian seamen lived oppressed lives, suffering severe discipline and harsh punishments, together with poor food, cramped living conditions, poor sanitation and primitive medical treatment. In reality this perception is not quite the case. Research reveals that common seamen were relatively accustomed to working and living in this environment and expected little else, providing they were treated with dignity and fairness.

The Impress Service was responsible for recruiting. It aimed to enlist skilled seamen

ABOVE Horatio Nelson as a young Post Captain in 1781. Promoted to this rank in 1779 at the age of 20, Nelson took command of the frigate HMS *Hinchinbrooke*. *(Historical Maritime Society)*

Life on the lower deck of a Georgian naval warship

ABOVE A Lieutenant relaxes in his cabin on the orlop of HMS *Trincomalee*. Lieutenants were in charge of deck watches and when a ship went into battle they might command a gun battery. *(Historical Maritime Society)*

Living conditions in the *Victory* were cramped. Some 480 men lived on the lower gun deck, sleeping in hammocks hanging from oak battens nailed to the overhead deck beams. With space being at a premium in the *Victory*, hammocks were spaced every 16in (40cm), but with half of the men on watch at any one time the overall space available would effectively increase.

Most of the seamen were berthed by divisions according to the 'part of ship' in which they worked. The marines and other men were berthed on the middle gun deck. Hammocks were taken down in the morning and stowed in nettings covered with tarpaulins that surrounded the open upper decks. These provided a barricade against small shot and musket fire in battle. Hammocks were not re-hung until 8:00pm.

Each man was allocated to a 'mess' where he ate at a table slung from the beams between the guns, or along the middle range of the deck to maximise the number of messes (there were some 60 to 80 tables needed). Seating comprised their sea chests; the stool benches often seen in paintings are a Victorian concept introduced for a peacetime navy.

Each mess comprised between six and eight men, who took it in turns to be the 'mess cook' and collect the victuals from the ship's Steward each day, prepare them and take them to the galley to be cooked by the ship's Cook. At meal times the 'mess cook' collected the food for his respective mess in a hook pot. Mealtimes – breakfast, dinner and supper – were at were at 8:00am, 12 noon and 4:00pm respectively.

Contrary to many common misconceptions, Georgian seamen generally faired far better than their civilian counterparts. Working a sailing man-of-war with little or no mechanical aids was a highly labour-intensive occupation, therefore the daily diet for each crew member needed to contain between 4,500 and 5,000 calories, the equivalent to that consumed by farm labourers, particularly during harvest season. The basic diet is shown in the box next page.

from the mercantile or fishing fleets, whose nautical experience was an asset. Although many men were 'pressed' into the Navy against their will in times of war, some 50 per cent were volunteers; and 25 per cent of those initially pressed were often rejected as unfit or unsuitable for the Service.

The major grievances that caused the great mutiny of 1797 among the ships of the Channel Fleet at Spithead concerned the fact that sailors' wages had not been increased since the time of Charles II, conditions of service being a secondary issue. Although officers were put ashore during this insurrection, order and discipline was maintained within the ships. It was the restrained and dignified action of the ships' crews that won the sympathy of the Admiralty and Government, leading them to address the shortcomings in pay and conditions.

Despite the hardships of naval life in men-of-war, the seamen retained loyalty to their ship. When faced with battle they fought more for their comrades in the ship and its commander, and for any potential prize money, than for any political agenda.

Standard rations per man

Day	Bread lbs	Beer pts	Beef lbs	Pork lbs	Pease pts	Oatmeal pts	Butter oz	Cheese oz
Sunday	1	8		1	½			
Monday	1	8				1	2	4
Tuesday	1	8	2					
Wednesday	1	8			½	1	2	4
Thursday	1	8		1	½			
Friday	1	8			½	1	2	4
Saturday	1	8	2					
Weekly total	7	56	4	2	2	3	6	12
Weekly (metric)	3.2kg	32ltr	1.8kg	0.9kg	1.1ltr	1.7ltr	0.8kg	0.34kg

Note: The above are the basic rations only and could be supplemented with raisins, vegetables, and with an equivalent amount of flour substituting bread. Although salted meat was carried, fresh meat and vegetables were obtained from friendly ports or victualling ships at every opportunity.

When at sea all the lower deck gun-port lids remained closed, which meant there was very little light or ventilation inside. Consequently the atmosphere was rank with stale sweat together with condensation and damp depending on weather conditions; small stoves were provided to combat this problem.

Toilet facilities comprised 'seats of ease' located on the beak deck, a short triangular platform overhanging the sea beyond the transverse beakhead bulkhead, closing off the fore part of the upper gun deck at the head of the ship – hence the term 'head' for toilet. Each mess had a piss bucket for use at night.

Seamen were categorised according to their ability and experience, and basically comprised three rates: ordinary seaman, able seaman, and petty officer. Among the latter were the Boatswain's Mates. These skilled men were mainly made up of those who worked aloft handling the sails and their associated ropes, knotting and splicing, operating the rigging and dealing with anchor and boat work.

Each man would be allocated to one of the ship's two watches, and further divided to work-specific tasks at each mast. Of these, the younger most nimble men, known as Topmen, worked precariously on the higher yards where agility combined with dexterity was essential. The older, sturdier seamen worked on the lower yards where strength rather than deftness was more vital, while those of long-standing and trustworthiness formed the afterguard working on the quarterdeck and mizzen mast. Irrespective of being a Topman, afterguard or otherwise, in battle most of these men would be working at the guns.

Most seamen earned about £1. 15s. 0d. per month, a wage that was comparable to a common labourer on the land. The inexperienced, called landsmen, provided the heavy labour for capstans and hauling on ropes until they attained sufficient skill to be rated ordinary seamen.

LEFT Bosun's Mates were Petty Officers promoted from within the ranks by the ship's Captain. They were responsible for the supervision of the seamen in their everyday duties and were answerable to the ship's Boatswain. Bosun's Mates were also responsible for dispensing punishment to the men in the form of flogging. *(Historical Maritime Society)*

Chapter Eight

Conserving and Restoring HMS *Victory*

───(●)───

HMS *Victory* was planned and built in the 18th century with an intended life expectancy of no more than 18 years. The fact that she exists today, almost 250 years later, is remarkable in itself. An ongoing programme of conservation will ensure that she lasts well into the 21st century.

OPPOSITE Mizzen top, fore top and main top stacked on the dockside for restoration. *(Jonathan Falconer)*

The problems

Supporting the ship

Regarding long-term conservation, the *Victory* is no longer in her natural environment wholly supported by seawater, for which she was designed. Instead, the ship is now permanently sitting in a dry dock supported by ten equally spaced steel cradles with the keel placed on a longitudinal concrete plinth. While this concept was considered the best practical solution when the ship was put into the dry dock in 1922, some 90 years standing in this state has now begun to prove detrimental to the well-being of the ship as a single artefact. Although initially thought reasonably supported, the cradles create adverse point loading on the wooden hull as most are not coincident with the main structural members of the ship. Furthermore, wood, a natural fibrous material with elasticity, does have an inherent tendency to 'creep' with age, insomuch as the ship is gradually sinking downwards under her own weight, creating adverse stresses upon the integrity of the ship's structure and fixing points. The only exception is the keel, which cannot

physically move downward providing the dry dock itself remains stable. Also, as the hull form distorts downward it is similarly spreading outward at its point of maximum breadth. Today, a sophisticated electronic system has been installed throughout the ship so that structural movement can be monitored in real time 24 hours/day and recorded for future analysis.

The elements

Because of her size the *Victory* cannot be protected within a case like a standard museum artefact. The expected museum experience surrounding conserved ships is for people actually to enter into and explore the vessel. Because the *Victory* is totally static to meet this accepted requirement and the method of support, the ship is constantly left to the mercy of the elements: rainwater, wind, sunlight and temperature fluctuations. Although it can be argued that it should be expected that the fabric of such ships should withstand these natural conditions without movement, i.e., rolling and pitching in a sea, these weather factors equally affect the general condition of the hull.

BELOW This is the stone dock in which the *Victory* is supported on steel cradles, set together with tubular steel breast shores giving transverse support direct to the dockside at the ship's greatest breadth.
(Jonathan Falconer)

Rainwater

Within the last decade the amount of rainwater that enters the ship has been significantly reduced by the introduction of modern sealing techniques in the upper deck. However, this remains a problem. No matter how much attention is given to sealing and caulking the hull and deck planking, rainwater continues to permeate down through some areas of the decks. This causes wet rot and decay in the many crevices formed within the complexity of the structure where rainwater can collect, and without natural hull movement this cannot be dislodged. Moreover, this endemic problem is further compounded by virtue of the fact that rainwater (which is effectively fresh water) leaches out the natural oils in the timber, reducing its natural defence against decay. Salt water, incidentally, does not readily bring about decay. The Danish preserved ship *Jylland* (1864), the world's last screw-propelled wooden steam frigate, is sprayed with seawater in the evenings to enable the salt water to soak into the weather-deck planking during the cool of the night. This practice allows the planks to swell and tighten up the caulked seams.

Wind

This element is not wholly detrimental to the hull fabric, but may cause deterioration in the paint finish. The fact that the ship cannot be physically turned around means that the starboard side of the *Victory* is constantly exposed to the prevailing south-westerly winds, which drive rainwater into the ship on this side.

Sunlight

Prolonged periods of very warm weather and sunshine can shrink the deck and ship's side planking and open up seams, with the result that following the next heavy rainfall rainwater penetrates the hull and causes the inherent problems discussed above. The American preserved frigate USS *Constitution* (1797), which remains afloat today, is completely turned around annually to avoid this effect and the wind problem.

Natural predatory attack

Made from a variety of timbers, the hull of the *Victory* is effectively a natural feast to insects and fungi. The most common insect problem relates to death-watch beetle, which attacks oak and lives in and bores through this timber to leave it weak and structurally wasted. This problem was found to be particularly predominant during the 1950s, but by introducing a programme to replace oak with alternative wood types where practical it has virtually been eliminated. Today, this is kept in check by constant monitoring and fumigation by pest control when outbreaks occur from time to time. Other insects attracted to the ship and eliminated by pest control are bees and wasps.

Fungal attack

Although occasionally found in the form of dry rot rather than wet rot, the most serious fungal problem relates to airborne spores that attack the wooden fabric. This has become more common with climatic change, winters becoming warmer and damper. These ideal conditions greatly encourage fungal growth with particular strains vigorously feeding off nutrients in the timber, rendering them weak and structurally wasted. This is not a modern-day problem for Samuel Pepys records that he found 'toadstools' growing in relatively new warships built during the late 17th century. While the fungi is easily removed where exposed, from a conservation perspective the roots of these growths can be well buried

BELOW Heel of the stern post and the inner post and copper fish plates supporting the tenons of the posts into the keel. Note the lower gudgeon brace for the rother (rudder) pintles. *(Peter Goodwin)*

like a cancer within the grain of the timber. Consequently where conservation policy allows, the most successful practice for eliminating this problem today is to cut back the surrounding material a suitable distance and replace with new timber.

Revisiting original construction techniques

Although every effort has been made to undertake reconstruction using original methods and materials, this approach has not always been practical because of a variety of limitations. These comprise the following:

1. Material choice.
2. Timber procurement.
3. Physical constraints.
4. Limited skill.
5. Cost.
6. Material processing.

Material choice

Although oak is the traditional building timber, unless it is well seasoned it can distort and is prone to shakes (cracks) along its grain. To avoid this today and the death-watch beetle problem, oak is very often substituted with alternative species of timber.

Timber procurement

Ignoring the species sought, all timber for *Victory* is carefully sourced following a strict code of practice. Finding legal timber of suitable size sometimes proves difficult and consequently this is governed by market availability. Very often this restriction necessitates obtaining timber of a lesser size and laminating it to bulk it out to the dimensions required.

Physical constraints

Reconstructing frames, re-planking and setting up pillars in the *Victory* is very labour intensive and today these tasks are governed by many Health and Safety issues. What was once acceptable traditional practice now has to be managed quite differently, applying modern aids such as chain blocks, cranes, jacks and workshop machinery facilities. Some hand-

crafting techniques are also now restricted.

Furthermore, to avoid lifting heavy planks up to 10in thick; the main wale for example, the strakes forming this planking are now built up in situ using a series of laminates to bulk it out to size. However, professional opinions vary on the suitability of lamination in some cases.

Limited skill

Today this presents a serious problem, primarily because traditional wooden boatbuilding and shipbuilding is being superseded by modern materials and construction and far fewer men or apprentices are available within the work pool. Equally, there are few shipwrights left who can shape timber with an adze and a side axe, traditional hand-held shipbuilding tools used since the time of early dynastic Egypt. Added to this, the naval dockyards, once the primary source for training in such skills, no longer provide apprenticeships in this field. However, within the private sector there are companies specialising in traditional timber ships who now actively run apprenticeship schemes in traditional shipbuilding.

BELOW In the dockyards of the 21st century craftsmen with traditional woodworking skills are hard to find. Here is a barrel stave being shaped by a carpenter on a cooper's horse.
(Colin Burring)

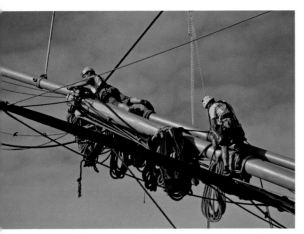

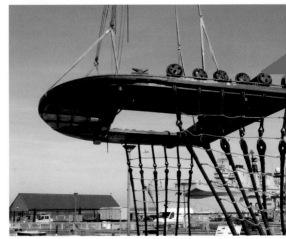

Cost

As with all restoration and conservation, financing ship-preservation projects, especially for historic ships made from wood, iron or steel, or a combination of each, is very difficult. Such projects incur significant costs both in terms of material procurement and the need to employ a suitably skilled workforce. Added to these constraints are the various hidden costs generated by the need for scientific investigation in order to effect quality conservation and protective techniques. Well-calculated compromises may have to be implemented which satisfy the available budget but are also in compliance with conservation policy for the ship. Historically, experience has shown that without a defined conservation policy and the application of a scientific approach to conservation and restoration, work undertaken over previous periods of the ship's life have already fallen short of predicted expectations to last well into the 21st century. This problem has been addressed through obtaining better materials, improved preservation techniques and condition-based maintenance regimes, although budgetary constraints mean that compromises sometimes have to be made.

Despite originally being planned and built with an intended life expectancy of no longer than 18 years (taking into account loss by inclement weather, accident or in battle), the fact that *Victory* still exists today is remarkable in itself. Conservation work that is funded and now under way will take *Victory* well into the 21st century with a significant reduction in the cost of ownership when the task is completed.

Material processing

Compared to the intensive labour required by the original builders, processing the material today for conserving the *Victory* is relatively easy using modern powered machinery and equipment, saws, drills, etc. Rather than being forged, fastenings are generally 'off the shelf

OPPOSITE In 2011 work began on a major restoration programme. The first task was to dismantle the ship's three masts, bowsprit and rigging. The last time the *Victory* was seen without her top masts was in 1944.
(MOD/Crown Copyright)

items', although ringbolts and eyebolts have been made within the dockyard.

Dockyard facilities and tools for the job

To reiterate, the dockyard generally provides well-equipped workshops for processing timber and producing wrought, forged or cast fastenings, As well as the modern powered machinery and equipment, portable powered tools are also available for use in the ship. Scaffolding is also erected to permit access to the work on the hull externally. Mast and rigging work is undertaken by skilled contract riggers who have the ability and expertise to work aloft, using harness climbing equipment and by means of mechanical high-rise platforms where necessary.

ABOVE The stern transom with all topmasts and topgallant masts struck, likewise the lower mast tops.
(Jonathan Falconer)

the frames. Time-consuming fabrication of smaller pieces of timber is commonplace and economical.

The shipwright's view – then and now

Since man first started making simple boats to work on the great rivers of early civilisation, e.g., the Nile, Tigris, Euphrates, Indus and Yangtze, and later more sophisticated sea-going trading ships, it inevitably became natural that those who constructed these vessels took great pride in their hand-crafted creations. This relationship between the shipwright and his ship extended when it became necessary to develop innovative purpose-built ships into complex war machines to protect trade; the Greek trireme is an early example.

Current romantic perceptions of how 18th-century shipwrights building warships viewed their work are quite wrong. In truth, building ships like the *Victory* was an extremely labour-intensive occupation. Not only were these men working in unprotected conditions and inclement weather, they were exposed to considerable hazards and as a result accidents were common. Furthermore, injury, muscular strain or illness would readily take its toll on individual men, who received little or no financial compensation to sustain them during periods of enforced absence. Also, it must be pointed out that for many of those physically building the *Victory* it was simply a skill and a job that earned them money to maintain their families. Whatever the long-term intended purpose of the ship they were working on, either for trade or war, the ship was only a tool with which these men were involved in creating.

The simplified view outlined above does not mean that the individual men involved in constructing the *Victory* did not uphold personal feelings and a sense of national pride about their work. We only have to consider the two key ceremonies relating to the ship, the laying of the keel and the final launching of the ship, to be reminded of this. These events were attended by both the workforce and dignitaries. Regarding pride, the year 1759 with its successive naval and military victories would have invoked considerable patriotism towards the King, the

Fabricating replacement parts

Today it is common practice to manufacture large components, such as knees, beams and riders, in laminated form to satisfy both strength and timber size limitations. Long baulks of teak planking are supplied made up in laminates in one plane and toothed bonded jointing in the longitudinal plane, the entire processes having been manufactured under controlled conditions including pressure bonding within the 'factories' of the timber merchants. Wood for timbers (ship's frames) is generally procured in rough-cut baulks that meet the desired scantlings, i.e., moulded breadth and siding (parallel) to

LEFT The head of ship showing un-planked hawse timbers and proceeding cant frames. *(iStock)*

country and its armed forces, especially as the recently penned words of Heart of Oak are our Ships was subconsciously uppermost in the minds of those working in the naval dockyards.

Note: the British naval march *Heart of Oak* was written by the composer William Boyce and given the lyrics by playwright David Garrick for his 1759 play *Harlequin's Invasion*.

In summary, the shipwrights who built the *Victory* and subsequently maintained her have in all probability accepted it as just part of their job, mainly as the *Victory* was for much of her time simply another part of the naval tool kit. However, the situation changed in the first

quarter of the 20th century when it was firmly decided to conserve the ship for posterity. The *Victory* is not used today as a warship in the true sense but is retained as a flagship. The shipwrights working to conserve the *Victory* now may not see the ship as their predecessors did, simply borne as a part of the war effort in times gone by. The new perception of current shipwrights is governed by their unique relationship with the *Victory*, a national icon of historical significance. More in favour of working with their hand tools, these shipwrights have chosen to stay and are so dedicated that some have reached their retirement having spent over 39 years working within this grand and most singular wooden ship HMS *Victory*.

List of sources and further reading

PRIMARY SOURCES

National Archives, Kew

TNA ADM 95/39 – Observations of the sailing qualities of His Majesty's Ship *Victory*, 1797.

TNA ADM 160 – Ship's ordnance records.

TNA ADM 180/10 – Progress books.

National Maritime Museum, London

NMM Letter books – Navy Board to Admiralty Vol. 24.

NMM No. 206B, Box 4, ZAZ 0-122 – *Victory* lines and draughts.

National Museum of the Royal Navy, Portsmouth

NMRN MSS1064/83.2376 – Record of Carpenter's and Boatswain's expenses for *Victory*, *Britannia* and *Africa*. July–December 1805.

NMRN MSS1998/41 – Journals of Master Gunner William Rivers in HMS *Victory*, 1793–1811.

NMRN MSS1986/573 – Journal of Midshipman William Rivers in HMS *Victory*.

Victory Archive, Portsmouth

VA VLD 2000 – Letters and documents.

SECONDARY SOURCES

Addis. C., 'The Men Who Fought with Nelson in HMS *Victory* at Trafalgar' (Nelson Society, 1988)

Anon, *Shipbuilder's Repository* (London, 1788)

Albion, R.G., *Forest Trees and Sea Power* (Harvard, Massachusetts, 1926)

Buglar. A.R., *HMS Victory: Building Restoration and Repair* (HMSO, 1966)

Callender, G., *The Story of HMS* Victory (London 1914)

Caruana, A., *A History of English Sea Ordnance 1523–1875, Vol. 2: The Age of the System 1714–1815* (Rotherfield , 1997)

Corbett J., *Signal and Instructions 1776–1794* (London, 1971)

Evelyn, J., *Sylva or a Discourse on Forest Trees* (1664, fourth edition, London, 1706)

Fenwick, K., *HMS* Victory (London, 1959)

Falconer, W., *Marine Dictionary* (London, 1769)

Fincham, J., *A History of Naval Architecture* (London, 1851)

Gardiner, R., *The Line of Battle* (London, 19??)

Goodwin, P., *Construction and Fitting of the Sailing Man of War 1650 to 1850* (London, 1987)

Goodwin, P., 'The Influence of Iron to Ship Construction: 1660–1830' in *The Proceedings of the 3rd International Conference of the Technical Aspects of the Preservation of Historic Vessels* (San Francisco, 1997)

Goodwin, P., 'The Hold Display of HMS *Victory*', *Model Shipwright* (London, 1996)

Gilkerson, W., *Borders Away, Vol. 1: Steel Edged Weapons and Pole Arms* (Lincoln, USA, 1991)

Gilkerson, W., *Borders Away, Vol. 2: Firearms in the Age of Fighting Sail* (Lincoln, USA, 1991)

Goodwin P. *Nelson's Ships: A History of the Vessels in Which He Served* (London, 2002)

Goodwin, P., 'The Construction and Development of HMS *Victory* and 18th Century Naval Warships' (Master of Philosophy Dissertation, Institute of Maritime Studies, University of St. Andrew's, 1998)

Goodwin, P., *Nelson's Men-O'-War: An illustrated History of Nelson's Navy* (London, 2003)

Goodwin, P., *Nelson's* Victory (London, 2004)

Goodwin, P., *Pitkin Guide Book of HMS* Victory (London, 2004)

Goodwin, P., *The Ships of Trafalgar: A History of the Ships, of the British French and Spanish Fleets 21st October 1805* (London, 2005)

Goodwin, P., 'Building the 100-Gun Ship Victory', *Transactions of the Naval Dockyard Society* (Portsmouth, 2011)

Goodwin, P., 'The Practice and Firepower of Broadside Firing in British Men of War in the 18th Century', *Shipwright*, (London, 2012)

Harland, J., *Seamanship in the Age of the Fighting Sail* (London, 1984)

Lavery, B., *Ship of The Line Vol. 1: Development of the Battlefleet 1650–1850* (London, 1983)

Lavery, B., *Ship of The Line Vol. 2: Design, Construction and Fittings* (London, 1983)

Lavery, B., *The Arming and Fitting of English Ships of War 1650–1850* (London, 1979)

Lavery, B., *Nelson's Navy* (London, 1989)

Lee. J., 'Interview with Peter Goodwin, Keeper and Curator HMS *Victory*', *Shipwright* (London, 2011)

McGowan, A., *HMS Victory: Her Construction, Career and Restoration* (London 1999)

McKay, J., *The Anatomy of 100-Gun Ship* Victory (London, 1987)

Rees, A., *Naval Architecture 1819–20* (Newton Abbot, 1970)

Scott, F., *The Square Rigger Handbook* (London, 1992)

Smyth, W.H., *The Sailor's Word Book* (London 1867)

Stilwell, A., *The Trafalgar Companion* (London, 2005)

Steel, D., *The Elements and Practice of Naval Architecture* (London, 1805)

Steel, D., *Elements of Mast Making, Sail Making and Rigging* (London 1794)

Steel, D., *Rigging and Seamanship*, 2 Vols (London, 1794)

ARTICLES IN THE MARINER'S MIRROR
(Journal of the Society for Nautical Research)

Goodwin, P., 'The Fore Topsail of HMS *Victory*', Vol. 83, No. 1 (London, 1997)

Goodwin, P., 'The Development of the Orlop deck in HMS *Victory*', Vol. 83, No. 4 (London, 1997)

Goodwin, P., 'The Influence of Iron to Ship Construction: 1660–1830', Vol. 84, No. 1 (London, 1998)

Goodwin, P., 'Where Nelson Died: An Historical Riddle Resolved by Archaeology', Vol. 85, No. 3 (London, 1999)

Whitlock, P., 'Jottings from HMS *Victory*', Vol. 62 (London, 1976)

Appendix 1

Glossary of terms

Abaft: Towards the stern.

Afore: Towards the bow or head.

Aft: Towards the stern or behind the stern.

Ahead: Towards or beyond the ship's bow.

A'lee: To be down wind.

Carling: Fore and aft set timber deck support fitted between beams.

Ledge: Short timber deck support fitted athwart ships between carlings.

A'luff: To be upwind.

Apron: Curved timber supporting the after face of the stem post.

Athwart: From one side of the ship to the other.

Astern: Toward or beyond the ship' stern. (See sternboard).

Bees: Flat wooden plates fitted at the bowsprit head to secure stays.

Bitts: Substantial vertical timbers (pins) fitted with cross pieces used for securing rigging (to belying pins) or for securing the anchor cables i.e. riding bitts.

Block: Wooden shell containing a pulley wheel providing mechanical advantage for hauling rigging.

Boatswain: Senior seaman warrant officer responsible to the ship's Master for rigging, sails, blocks anchors, ground tackle ship's boats, cables and all their related stores and equipment; likewise the organisation and training of the seamen.

Boltrope: Rope stitched around the edges of a sail to provide stiffening.

Bowline: (1) Form of knot that does not slip. (2) Rope attached to the leach of a sail to hold the leach to windward.

Breasthook: Large internal wooden timber fitted horizontally to stiffen the frame structure at the fore end of the ship bracing the hawse pieces.

Breastrail: Safety rail crossing the end of the forecastle, quarterdeck or poop deck.

Bunt: Bottom edge of a square sail. Term also used for the centre of a yard.

Buntline: Running rigging used to haul up the bunt of a sail to the yard when taking in and stowing sail on the yard.

Cant frame: Ship's timber or frame fitted out of square from the keel where the hull shape commences to narrow towards the bow or towards the stern at the wing transom.

Cap: Rectangular wooden block fitted at the lower mast head, topmast head or bowsprit to house the respective mast above it.

Carriage: Wooden mounting or support for a gun.

Case of wood: Cylindrical wooden container made from poplar or willow, with an elm lid, used for carrying gun powder cartridges from the magazines to the gun decks.

Ceiling (Sealing): Inner planking sealing the inside of the lower part of the ship laid up between the strakes of thickstuff.

Chain plate: Wide horizontal wooden platform projecting from the ship's side to provide greater spread for the lower rigging. The shrouds for the masts are secured to the deadeyes mounted on the chain plate.

Chains: See above.

Channel or Channel Plate: See chain plate.

Clamp: Band of heavy longitudinal planking fitted to the insides of the ship's frames to support the ends of the deck beams.

Cleat: Pieces of timber shaped with two horns for belaying ropes to.

Clew: The lower outer corners of square sails and aftermost lower corner of triangular sail.

Clewline (or Clew-garnet): Rigging used with a garnet block to haul up and in the clew of a sail.

Counter timber: Curved timber projecting upwards and aft to form the shape of the lower part of the stern transom.

Coxswain: Petty Officer responsible for steering (the origin is Anglo-Saxon, from *Cog's Swain*).

Crosstrees: Set of fore and aft and transverse timbers fitted at the head of a topmast to support the heel of the topgallant mast; used as a lookout point.

Crutch: Curved wooden timber fitted internally to brace the framing of the ship towards the stern or bow one housing the heel of the foremast, another, the mizzen mast heel.

Deadeye: Wooden disc made from *lignum vitae* manufactured with grove around its periphery to receive the lower end of a shroud, standing rope or iron band It is and axially pierced with three holes through which lanyard is laced to its lower (standing) counterpart.

Deck hook: Transverse timber supporting the foremost or after end of a deck.

Drumhead: Upper section of a capstan that receives capstan bars.

Earring: Rope eye formed in the bolt rope of a sail at its uppermost outer corner.

False keel: one or two layers of sacrificial timber fitted under the main keel before launch by means of copper staples for easy removal.

Fife rail: Fifth or highest rail running longitudinally along the ship's side.

Figure: Formal term for the carved ornamentation fitted at the head of the ship. Taking the form of either human, animal or bird, likewise a coat of arms, badge or fiddlehead. This feature is colloquially called a figurehead.

Flat: Planking forming the 'flat of the deck'.

Floor: Lowermost timber crossing the keel forming part of a ship's frame (or timber).

Fore: Towards the bow or head of the ship.

Futtock: Curved timber forming the shape of a ship's frame (or timber) in the lower body of the ship's hull (usually four are used).

Futtock shroud: Shrouds set at an angle leading up and outward towards the outer edge of a top providing both continuous climbing access onto the top (or crosstrees) and forming an anchor point for the topmast shrouds.

Hair bracket: Ornamentally molded serpentine shaped timber running from the head rail to the back of the 'figure'.

Halliard: (Origin *haulyard*). Running rigging for hoisting yards, staysails, booms etc.

Hanging knee: Wooden bracket set in the vertical plane connecting the end of a deck beam to the ships side; its vertical arm bolted to the inner face of the ship's side, the horizontal arm bolted to one side of the beam.

Hawse piece: Timbers or frames forming the shape of the ship's bow so called as they are pierced to provide a hawse (Old English/Anglo-Saxon for throat) hole for passing the anchor cable.

Heave to: To temporarily stop the ship by setting the sails to counteract the wind effect upon opposing sails to prevent the ship standing into danger or taking in or lowering boats etc.

Hook and butt: System of scarphing employed for interlocking planking used for the beam shelves, deck clamps, spirketting, thickstuff and wales.

In irons: A ship is said to be 'in irons' (clamped or restrained) when unable to turn her head through the wind when tacking etc. The alternative expression is 'missed stays'.

Jeer block: Large double block with two or three shivers (sheaves) used for hoisting the fore or main yard.

Keelson: Innermost longitudinal keel made from of oak or elm oak laid over the floor timbers and crosspieces to lock them in position.

Kevel: Vertical post fitted against a bulwark furnished with a shiver for running rigging i.e. topsail yard tie halyards. The head of the post is formed timberfashion for belaying rope.

Knighthead: The upper ends of the innermost hawse timbers that embrace the bowsprit.

Knee of the head: Curved shaped structure fitted to fore face of stem post supporting the head rails and 'figure'.

Larboard: The left hand side of the ship as you face towards the bow. The name originates from Old English, '*laden board*', the side of the ship which was brought alongside a wharf to embark (laden = load) or disembark cargo etc. This side was used to prevent damage to the steer board on the opposite side of the ship (see *starboard*). The term larboard was generally superseded by the more commonly known term Port (the side of the port the ship is visiting or side of the access port used the embarkation) The term port was introduced with the change to navigation rules in 1844 to avoid the word *larboard* being confused by the word *starboard*, i.e. during noisy gales and original helm orders.

Leach: The vertical side of a square sail, or after edge of a triangular jib or staysail, likewise the mizzen sail or spanker.

Lee or Leeward: Side of the ship downwind or 'a'lee' of the prevailing wind.

Lengthening piece: Vertical timber frame forming the ship's side framing above the waterline extending from the gun port lintels or toptimbers to the plank sheer at the top of the side.

Light: Glazed window, i.e. the stern lights or for illuminating magazines and storerooms below the waterline.

Limber boards: Short potable boards with hand holes made of elm board laid at an angle the between side of the keelson and rabetted edge of the limber strake forming a longitudinal cover over the limber passages.

Lining: Inner planking wrought longitudinally between the gun ports above the spirketting below the deck clamps (or beam shelves). Easy to apply it is colloquially referred to as the 'quickwork'.

Lintel: Horizontal timber checked into the vertical faces of ship's frames forming upper boundary of a gun port.

Lodging knee: wooden bracket set in the horizontal plane connecting the end of a deck beam to the ships side; its long arm bolted to the ships side the short arm bolted to the beam.

Luff: Fore edge of a triangular sail, i.e. jib or staysail or fore edge of the mizzen sail or spanker.

Luff: To luff is to turn into the wind setting sails aback to stop making headway.

Oakum: Old rope fibre mixed with tallow and tar to caulk seams of deck or hull planking.

Orlop: (Origin Old German.) A flat mezzanine fitted below the lower gundeck, 'overlapping the hold with planks '. The amidships section comprises short length loose fitted oak boards 1½in thick let down onto rabetted orlop beams. The foremost and aftermost area sections of the comprise fixed planking forming a protective cover over flammable storerooms and explosive magazines below.

Poop: Short-length deck forming cover over the cabins at the after end of the quarterdeck, originating from the Latin *puppis* meaning aftermost deck. The poop is used as the signal deck.

Quick work: See Lining.

Quoin: Wedge shaped oak block set on gun bed used for adjusting angle of a gun.

Robbands: Ropes used to lace the head of a square sail to its respective yard or gaff boom.

Scarph: Method of joining two lengths of timber together using opposing flat surfaces or a hook joint.

Scupper: Lead pipe fitted to drain water from decks or pumps overboard or a longitudinal channel running along the outboard edge of a deck at the waterways to convey water.

Shell: Hollow iron spherical explosive projectile filled with gunpowder ignited with fuse when fired from gun.

Sheet: Running rigging rope passing from the clew of a sail used to for setting a sail down and outward, i.e. the sail is sheeted home and set.

Shot: Solid iron spherical projectile fired from guns to damage opponent ships or fortifications. Referred to as round shot. It does not explode. (See Shell).

Shot comprises the following types:

1. Bar: iron bar with either spherical round shot, solid iron discs (hammerhead) fitted at each end to damage rigging masts and spars.
2. Canister: anti-personnel weapon with a shot gun effect comprising a tin canister containing round shot or large quantity of musket balls.
3. Chain: chain with spherical round shot at each end.

4 Elongating: similar to bar shot with interlocking bars that extend its length when spinning through the air.

5 Grape: anti-personnel weapon with a shotgun effect comprising a flat iron plate and central pin mounting small solid round shot contained in a quilted canvas bag the latter of which disintegrates when fired; that for 12-pounder gun comprises 9 x 2 lb balls; a 24-pounder 9 x 4lb balls and a 32-pounder 9 x 3lb balls.

6 Langrage: anti-personnel weapon with a shotgun effect comprising canvas bag containing junk iron nut and bolts roves etc., or stone pebbles.

Shot rack: Wooden plank fashioned to hold ready-use spherical shot fitted upon hatchway coamings or at the bulwarks.

Sill: (Cill) Horizontal timber checked into the vertical faces of ship's frames forming lower boundary of a gun port.

Sleeper: Large curved timber fitted internally at the after end of the hold to provide longitudinal support to the transom beams below the wing transom

Slings: (1) Rope support strop passing round the middle of a yard and over the mast head cap. These were doubled up in battle with preventers made of chain. (2) Term used for the centre section of a yard.

Spirketting: Internal band of thick planking wrought between the deck waterway and sills of the gun ports providing longitudinal strength to counteract the hogging and sagging effects upon the ship's hull.

Sponge: Wooden cylindrical block coated with sheep skin mounted on an ash stave used with water for extinguishing burning debris in the bore of a gun before loading.

Square sail: Quadrilateral sail bent (laced) onto a horizontal spar called a yard.

Standard: Inverted timber knee used to support a beam (i.e. orlop) where a hanging knee of less than 90 degrees is structurally unsupportive due to the inward curve of the ship's side towards the keel.

Starboard: The right hand side of the ship as you face towards the bow. The name originates from the Old English '*steer board*' used before the innovative introduction of a central rudder.

Staysail: Triangular or quadrilateral sail set on a stay or staysail stay between masts set in light winds or to steady the ship in a rolling sea.

Steering sail: Square sail bent to its own yard set up on boom extending from the yard arms set to provide greater speed in light winds

Sternboard: The movement of the ship being driven astern or backwards.

Sternson: Timber knee giving fore and aft support to the fore face of the sternpost and inner post often formed as an extension to the keelson.

Stunsail: (See Steering Sail).

Tack: Rope secured to the clue of a sail to haul the foot of the sail forwards or to one side in the case of jibs and staysails.

Tacking: (1) Sailing manoeuvre made to turn the ship head through the wind to lay the ship on an alternative tack or course. (2) Tacking to drive the ship on a series of diagonal courses when sailing against the wind.

Taffrail: (Taferal) Ornate part of the ship's stern above the stern galleries.

Tie: (Tye) Runner and tackle for the halyards to hoist topsail or topgalant yards.

Thickstuff: Band of heavy inner planking wrought longitudinally over the head and heel scarphs (wrungheads) of floors and futtocks.

Thimble: Wooden truck with an axial hole to pass a light hauling rope. The periphery has a groove to fasten a lizard (strop) Use: brails, topgallant, or royal clewlines and buntlines.

Timberhead: Upper end of a ship's frame timber formed into a bollard for belaying.

Top and butt: System of tapered planking employed for the ship's wales, maximising timber use from the bowl of a tree without waste.

Toptimber: Vertical timber frame forming the ship's side framing above the water line extending from the 4th futtock head terminating at gun port sills or plank sheer at the top of the side.

Transom: (1) Collective term for the ship's stern above the waterline comprising upper and lower counters stern galleries, lights, cove and taffrail. (2) Flat stern of a ship's boat.

Transom beam: Shaped horizontal timbers bolted to the back of the stern post, the outer ends scarphed into fastened piece (aftermost square frame).

Tricing line: Line used to 'trice in' (haul or hoist) a tackle, i.e. yard arm tackle.

Truck: Flat wooden disc made from elm used for (1) gun carriage wheels. (2) Stopper at the pole end of a spar, i.e. topgallants and royals mast, flying jib boom, ensign and jack staffs.

Trundlehead: Lower section of a capstan that receives capstan bars.

Up and Down: Term used to inform the Master that the anchor cable is vertical indicating that one final heave on the capstan is required to break the anchor from the ground (seabed) and prelude to making sail.

Wadhook: Iron 'corkscrew' shaped tool mounted on an ash stave for extracting wads, debris and misfired cartridges from the bore of a gun.

Wale: External band of thick planking binding the ship's frames together providing longitudinal strength to counteract the hogging and sagging effects upon the ship's hull.

Wear: Sailing manoeuvre made to turn the ship stern through the wind to lay the ship on an alternative tack or course.

Whelp: Thick vertical timbers fitted to maximize the working diameter of a capstan.

Windward: (1) The side of the ship from which the prevailing wind is directed. (2) To be to windward is to be upwind (see A'luff).

Wing transom: Large transverse timber set down upon the head of stern post on which the stern counter timbers and entire structure above is supported.

Yard: Horizontal spar suspended from a mast on which a sail is bent.

Yardarm: The outer extremity of a yard beyond the rigging cleats.

Appendix 2
Essential dimensions, weights, etc. for HMS *Victory*

Class name	Victory
Rate	First
Design date	1759
Designer	Sir Thomas Slade
Builder	John Lock/Edward Allin
Yard	Chatham Dockyard
Ordered	24 June 1759
Keel laid	23 July 1759
Launched	7 May 1765
Length of gun deck	186ft
Length of keel for tonnage	151ft 3⅝in
Extreme breadth	51ft 10in
Depth in hold*	21ft 6in
Tons burthen	2162.22/94
Complement	850
Total armament	100
Lower gun deck	30 x 42-pounders
Middle gun deck	28 x 24-pounders
Upper gun deck	30 x 12-pounders
Quarterdeck	10 x 6-pounders
Forecastle	2 x 6-pounders

* This measurement is taken from the underside of the lower gun-deck planking to the upper side of the limber strake next to the keelson and consequently includes the 6ft 6in height of the orlop. Thus, the true height of the hold is actually 15ft.

Source: Goodwin, P., *Nelson's Ships: a history of the vessels in which he served 1771–1805* (London 2002), p234. Also Lyon, D.,*The Sailing Navy List* (1993), p231.

Appendix 3
HMS *Victory* visitor information

Location
HMS *Victory* is on public display in No. 2 Dry Dock at Portsmouth's Historic Dockyard in Hampshire, Great Britain. The Victory Gate for the Historic Dockyard is in Queens Street by Portsmouth Hard.

Opening Hours
The Historic Dockyard opens at 10:00 everyday throughout the year, except on Christmas Eve, Christmas Day and Boxing Day when it is closed.

From April–October: Last tickets to the attractions are sold at 1630; last entry to HMS *Victory* is at 16:45 and the Dockyard gates are closed at 18:00.

From November–March: Last entry to Victory is 16:00 and the Dockyard gates are closed at 17:30.

HMS *Victory* closures
HMS *Victory* remains a commissioned ship of the Royal Navy and the flag ship of the Second Sea Lord and Commander-in-Chief Naval Home Command. Due to service commitments, there are occasions when *Victory* will be closed to the public for all or part of the day. Further details of planned closures can be found at http://www.hms-victory.com

Ticket prices
There are a range of ticket options available for HMS *Victory*. Your ticket for the ship includes entry to the other attractions on the Historic Dockyard site. For details of ticket prices visit the Portsmouth Historic Dockyard website at http://www.historicdockyard.co.uk

Index

20th Century Fox 80

Action damage repair 118
Addington, Henry 31
Admiral's quarters 10, 52, 54-56, 108-109
Afterguard 157
Allin, Edward, Master Shipwright 24-25
American Declaration of Independence 18
American War of Independence 28-29, 87
Anchor deck 57
Anchors and ground tackle 20, 68-72, 74-75, 111
Anson, Admiral George 14
Apron 20, 24
Armament 80
 ammunition 98-99
 canister shot 96
 grappling irons 101
 grenades 97, 101
 small arms 100
Atkinson, Thomas 68

Balchin, Admiral Sir John 10
Ballast 40-41, 148
Baltic and Peninsular Wars 35
Barrel rooms 46
Bazely, John 159
Beak deck 56, 59, 161
Beam shelves 24
Belfry 56-57
Binnacle 67
Booms 60, 62, 110, 112
Bowsprit 20, 53, 57, 150, 169
Boyce, William 171
Bread room 22, 40, 47, 49
Breast hook 24
Brest blockade 32
Bulkheads 42, 108
 beakhead 25, 35, 52, 58-59, 108
 main transverse 54
Bulwarks 31-32, 83, 108

Cabins 10, 46-48, 54, 58, 109, 156
Cables 47, 49, 70-71, 73-75, 118
 maintenance 122
 messenger (vyol) 73-75, 118-119
 refitting at sea 122-123
Cape St Vincent, Battle of 30
Capstans 48, 71-75, 110, 118-123, 161
 crabs 22
Captain's quarters 47, 58, 108
Captain's role and qualifications 159
Carpenter 47, 110, 114, 118
Cartridge racks 44
Caulking 105-107, 165
Chaplain 49

Chatham Dockyard 9-10, 14, 16-17, 23, 25, 30-31, 35
Coal hole 40-41, 47
Cockpits 47-48
Collingwood, Admiral Cuthbert 32-33
Compasses 58, 67, 148
Complement of personnel 156, 160-161
 able seamen 161
 officers 49, 159-160
 ordinary seamen 48, 161
 petty officers 161
Construction 16-19, 23-24, 42, 167
Cooper, Captain Thomas 16
Copenhagen, Battle of 31
Cordoba, Admiral Don José 30
Cornwallis, Admiral 32
Corsica, Invasion of 29-30
Costs of construction and refits 9, 25, 28-29, 31, 35

Daily routine at sea 158
Deadeyes 39, 60, 62, 105, 113
Deane, Anthony 9
Death-watch beetles 36
Deck beams and planks 24, 165
Decks and internal arrangements 40-59, 103
Deptford Dockyard 14
Design and development 9-10, 13-17
Dockyard maintenance and refitting 104-106
Drake, Admiral 28
d'Orvilleir, 28

Fastenings (bolts nails, roves and spikes) 18, 22, 28, 39, 41, 170
Figureheads 5, 23, 31
Filling room 41, 44-45
Firepower 80
First World War 36
Fish room 40, 47
Flags and lockers 58-59
Floors 23
Forecastle 31, 53, 57, 62, 71, 75, 80, 85, 105
Forecastle-men 157
Fore stays 113
Forge, portable 128-129
Frames 21-25, 170
 cant 22, 171
 green 24
French Revolutionary War 29, 31, 87, 101
Futtocks 20-24, 170

Galley 53, 56-57
 pantry and cupboards 51
 stove 51
Galveston Historical Foundation 140
Garrick, David 171
Geary, Admiral 28

Gibraltar 28, 33, 35
Gun batteries 9
Gun crews 156
Gun decks 19, 79-80, 156, 159
 lower 28, 48-49, 68, 73, 75, 104, 156, 160-161, 170
 middle 10-11, 20, 24, 50, 52-53, 72, 79, 81, 105, 156, 160
 upper 10, 29, 52, 56-57, 83, 105, 160
Gun equipment 86-88
 muzzle lashings 95
 ropes and cordage 82, 94-95, 112
 salt box 86, 88
 side arms 87
 tompions 88
Gun ports and lids 20, 24, 29, 31-32, 95, 105, 108, 111, 161, 170
Gunpowder 41-42, 44-45, 96
Gun room 22, 49-50, 68, 156
Guns and gun carriages 8, 15, 20, 79-93, 98
 Armstrong pattern 87
 Blomefield pattern 87
 drill 90-93
 embarking 150-151
 firing 79, 89
 maintaining 115-118
 recoil 94, 112
 velocity 85
 6-pounder 28
 12-pounder 28, 44, 47, 52, 57-58, 62, 80-81, 83-86, 94, 150
 18-pounder 32, 80, 84
 24-pounder 10-11, 44, 47, 50, 77, 80-83, 85-86, 93-94, 108, 112, 116
 32-pounder 28, 31, 45, 80, 82, 85, 94, 150
 42-pounder 28, 31, 82
 64-pounder 57
 68-pounder 56, 80, 85

Hamilton, Emma Lady 55
Hammocks 10, 48-50, 156, 160
Handling 132
Hardy, Admiral Thomas Masterman 28, 31-33, 36, 159
Harlequin's Invasion play 171
Hawse pieces and holes 20, 70, 171
Hawke, Admiral 10
Heart of Oak naval march 171
HMS abbreviation 8
Hogging 25, 106
Hold 40, 47-48, 118
Hood, Admiral Samuel 29-30
Hotham, Admiral William 30
Howe, Admiral Richard 28-29
Hull 13-14, 17-19, 22, 24-25, 28-29, 31, 39, 42-43, 103, 164-165, 169
 copper sheathing 28, 36, 166
 maintenance at sea 104
 supporting in dry dock 164, 166

Impress Service 160
Inner post 19
Ironwork 18

Jervis, Admiral Sir John 30

Kariskrona 35
Keel 16, 18-19, 22, 164-165
 false 166
 laying 16, 170
Keelson 22-24
Kempenfelt, Admiral Richard 10, 28-29
Keppel, Admiral Augustus 28
King Charles I 9
King George I 10
King George III 81
King Henry VIII 118
King James II 9-10
King Louis XVI
Knees 9, 35, 59, 104, 170
 dead wood 19
 standards 22
Knight head timbers 20
Kronstad blockade 35

Laid up 28, 35
Landsmen 44, 157, 161
Launching 7-8, 25, 28, 170
Life expectancy 163, 169
Light boxes and light rooms 45-47
Linzee, Admiral Robert 30
Litharge 107
Living conditions 11, 48, 156, 160-161
Lock, John, Master Shipwright 16, 24
Lock Snr, John, Master Shipwright 16
Lock, Pierson, S Master Shipwright 16
Log books and journals 104, 110, 146
Lucas, Captain Jean Jacques 34

Magazines 40-41, 43-44, 46-47
Man, Admiral Robert 30
Manger 48, 70-72
Manoeuvrability 14, 118
Manoeuvring
 bringing to anchor 146
 getting under sail from anchor 142, 144-145
 heaving to 140
 tacking 136-137, 147
 wearing 131, 138-139
Manpower required 74
Marines' walk 57
Marryat, Captain 157
Master and Commander film 80
Mast room 47

Masts 31, 60, 62, 103-104, 109, 112, 169
  bracing 112
  broken 110
  fore lower 60
  foremast 48, 52-53, 133
  foretopgallant 114
  lower 18, 113
  main 40, 50, 52-53, 58, 60, 63, 157
  mizzen 25, 34, 47-48, 50, 55, 58, 157, 161
  replacement 60
  sending down 112-113
  topgallant 25, 112, 113
  topmast 25, 34, 112-113, 148
Materials 39, 103-104, 118, 167, 169
Messes 10, 40, 48-49, 160-161
Moore, General Sir John 35

Napoleon Bonaparte 31-34
Napoleonic Wars 31, 87, 101
National Archives 104
National Maritime Museum 23
National Museum of the Royal Navy, Portsmouth 23, 87, 113
Navy Board minutes 15
Nelson, Admiral Horatio Lord 7-8, 28, 30, 32-34, 80, 159
  fatally wounded 33-34
  knighthood 30
  loss of sight in right eye 30
Nielson, Tommi 107

Oakum 105-107
Ocean Institute, California 137
Odyssey Marine Exploration 10
Officers' accommodation 50, 156
Orlop 29, 46-48, 97, 109, 118, 156

Painting 28-29, 31, 35, 106-109, 114-115, 165
Pallating flat 41-42, 44-45
Parker, Admiral Sir Hyde 28, 31
Peace of Paris 23
Pepys, Samuel 165
Performance 152-153
  speed 14, 28, 118, 152
Pett, Phineas 9
Pitt, PM William the Elder 14, 16, 25
Pitt, PM William the Younger 31
Planking 22, 24
Plymouth Dockyard 16
Poop deck 32, 58-59, 105, 108, 157
Portable bulkheads 10, 50, 53, 55, 58
Portsmouth Historic Dockyard 7-8, 16
Portsmouth Dockyard 9-10, 30, 35-36
Powder room 47
Preservation
  national appeal 36
  opened to the public 36
  placed in dry dock 7, 36, 164
Provisions 47
  embarking 149
  replenishment at sea 149

Pulley blocks 49, 62, 132
Pump wells 40, 47, 127
Pumps 40, 118, 123-128
  chain (yard) 48-49, 123, 125-128
  elm tree 40, 53, 123-125, 127
  maintenance 124, 126

Quarterdeck 28, 31, 33, 57-58, 66, 68, 79-80, 84, 105, 108, 157, 161
Quarter galleries 25, 59
Quiberon Bay, Battle of 10

Racking 104
Rammed accidentally 36
Rations 161
Recommissioned 32, 35
Red-hot shot 129
Refits and rebuilds 174; 1783 28; 1787 29; 1794 30; 1800-03 23, 31, 68; 1814-16 18, 35; 1888 36; 2011 169
Restoration and conservation 10, 36, 118, 163-171
  costs 169
  dockyard facilities 169
  fungal attack 165, 167
  health and safety issues 167
  natural predatory attack 165
  rainwater 165
  replacement parts 170
  sunlight 165
  tools 169
  wind 165
Rigging 57-58, 62-64, 103, 110, 112-113, 169
  blacking down 113
  maintenance at sea 112
  repairs to action and storm damage 113
  running 112
  standing 112-113
Roles
  flagship 7-8, 27-30, 32, 35-36, 80, 108
  in use as a tender 36
  living museum 8
  use as a troopship 35
Ropes 50, 55, 62, 69, 94-95, 112, 132, 134, 161
  breeching 82, 94, 112
Rot, wet and dry 36
Royal Armouries Fort Nelson 87
Rudder (rother) 18, 50, 66, 69-70, 165
  loss of 69

Sagging 106
Sailcloth 113
Sailing qualities 14-15
Sail room and storage 46-47, 65-66
Sails 15, 57-58, 65, 103, 110
  foretopsail 113-114
  headsails 150
  main 157
  maintenance at sea 112-113
  mizzen 31, 135-136, 148, 157
  moving and hoisting 112
  reefing 141

royals 145, 147
  setting 132-133
  spritsail 150
  square 134
  staysails and jib sails 135-137, 143, 147-148
  topgallant 137, 141, 145, 147-148
  topsail 136-137, 140-141, 143, 145, 147-148
  wind pressure 110
Samaurez, Admiral Sir James 35
San Fiorenzo, Calvi and Bastia, Siege of 30
Sea trials 28
Seaman's view 160
Seams 104-107
Seppings, Sir Robert 24, 35
Seven Years War 14, 17, 23
Sheerness Dockyard 24
Ships
  Acheron 80
  Achille 34
  Agamemnon 30
  Alert 159
  Amphion 32-33, 159
  Bounty 145
  Captain 30
  Constitution 132, 165
  Duke of Wellington 36
  Elissa 140, 157
  Endeavour 132, 134, 141, 148, 150
  Impregnable 27, 30
  Invincible 101
  Jylland 165
  La Bucenture 34, 80
  Lexington 159
  Martin 36
  Mary Rose 118
  Neptune 34-36
  Pilgrim 133, 137, 147
  Prince (Royal William) 9
  Prince Royal 9
  Rédoubtable 34
  Royal George (f. Royal Anne) 10, 29
  Royal James 9
  Royal James (Victory) 10
  San Josef 30
  San Nicholas 30
  Sovereign of the Seas (Royal Sovereign) 9, 34
  Shah 60
  Stat Amsterdam 143
  Surprise 80, 131, 143
  Trincomalee 132
  Unicorn 132
  Victory (Balchin's) 10, 23
  Warrior 132
Ship's boats 15, 76-77, 80
  hoisting in and getting out 151, 157
  maintenance 114-115
Ship's time 159
Shipworm 28
Shipwrights 20, 24, 103, 167, 170-171
Shot lockers 31, 40-41

Shrouds 60, 62, 112-113
Sick berth 52
Slade, Sir Thomas 13-14
Society for Nautical Research 36
Solebay, Battle of 10
Spirit room 40, 47
Staffs 57, 62
  jack 57
Standing 'in frame' 23
Steerage 55
Steering arrangement 66-67
  failure 68-69
Steering wheel 49, 55, 58, 66-69
Stem post 19, 166
Stern 19, 31, 108-110
  decoration 35, 109-110
  gallery and lanterns 10
Stern post 18-19, 66, 70, 165
Sternson 19, 22
Stockholm tar 105-106
Storage and storerooms 15, 40-41, 46-49
  gunners' 97-100
Stores and equipment, embarking 148
Surgeon's dispensary 48
Sutton, Captain Samuel 32-33

Tackle 94-95, 112
Tiller 49-50, 52, 55, 66-69
Timber species 17-18, 39, 60, 165, 167
Timbers 19-20, 24, 164-167, 170
Toilets (heads, seats of ease) 52, 54, 56, 58-59, 161
Topmen 157, 161
Toulon, Siege of 29-30, 32-33
Trafalgar, Battle of 7-8, 27-28, 31-34, 36, 68-69, 79-80, 113-114
  centennial celebrations 36
Transom beams 19
Tumblehome 96, 108

Ushant, Battle of 28-29

Ventilation 104, 161
Victory name 8, 14, 23, 108, 110
Villenueve, Admiral Pierre 33-34
Voice tube 31, 68

Waist 57, 105
Waisters 157
Wales 24-25, 108, 167
Wardroom 50, 156
Watch system 45, 156, 158
Watering 149
Waterloo, Battle of 80
Wellesley, General Sir Arthur (Duke of Wellington) 35, 80
Whitewash 109
Wilson, Robert 157
Wing transom 19, 22
Woolwich Dockyard 9-10, 14

Yards 57, 60, 62-63, 103-104, 109-110, 112, 133, 136, 140, 157
Yorke, Admiral Sir Joseph 35

178